Abeer Alarawi
Iman Roqan

Europium Doped Zinc Oxide Nanorods, Structural and Optical Properties

AF138227

Abeer Alarawi
Iman Roqan

Europium Doped Zinc Oxide Nanorods, Structural and Optical Properties

LAP LAMBERT Academic Publishing

Impressum / Imprint

Bibliografische Information der Deutschen Nationalbibliothek: Die Deutsche Nationalbibliothek verzeichnet diese Publikation in der Deutschen Nationalbibliografie; detaillierte bibliografische Daten sind im Internet über http://dnb.d-nb.de abrufbar.
Alle in diesem Buch genannten Marken und Produktnamen unterliegen warenzeichen-, marken- oder patentrechtlichem Schutz bzw. sind Warenzeichen oder eingetragene Warenzeichen der jeweiligen Inhaber. Die Wiedergabe von Marken, Produktnamen, Gebrauchsnamen, Handelsnamen, Warenbezeichnungen u.s.w. in diesem Werk berechtigt auch ohne besondere Kennzeichnung nicht zu der Annahme, dass solche Namen im Sinne der Warenzeichen- und Markenschutzgesetzgebung als frei zu betrachten wären und daher von jedermann benutzt werden dürften.

Bibliographic information published by the Deutsche Nationalbibliothek: The Deutsche Nationalbibliothek lists this publication in the Deutsche Nationalbibliografie; detailed bibliographic data are available in the Internet at http://dnb.d-nb.de.
Any brand names and product names mentioned in this book are subject to trademark, brand or patent protection and are trademarks or registered trademarks of their respective holders. The use of brand names, product names, common names, trade names, product descriptions etc. even without a particular marking in this works is in no way to be construed to mean that such names may be regarded as unrestricted in respect of trademark and brand protection legislation and could thus be used by anyone.

Coverbild / Cover image: www.ingimage.com

Verlag / Publisher:
LAP LAMBERT Academic Publishing
ist ein Imprint der / is a trademark of
OmniScriptum GmbH & Co. KG
Heinrich-Böcking-Str. 6-8, 66121 Saarbrücken, Deutschland / Germany
Email: info@lap-publishing.com

Herstellung: siehe letzte Seite /
Printed at: see last page
ISBN: 978-3-659-60673-1

Zugl. / Approved by: Thuwal, King Abdullah University of Science and Technology, Kingdom of Saudi Arabia, 2014

Copyright © 2014 OmniScriptum GmbH & Co. KG
Alle Rechte vorbehalten. / All rights reserved. Saarbrücken 2014

Structural and Optical Properties of

Eu Doped ZnO Nanorods prepared

By Pulsed Laser Deposition

By Abeer Alarawi.

Dr. Iman Roqan.

1

Acknowledgment

I owe my every achievement to Allah,

My Kids (Melaph, Mayar and Yamin),

My Parents (Dad, Mom),

My Husband Nayef,

And my sister Nada,

And this is where I dedicate the BOOK to all of them...

TABLE OF CONTENTS

LIST OF FIGURES

LIST OF TABLES

Chapter 1

Introduction

ZnO is an II–VI compound that has been studied since 1935. ZnO is a direct and wide-band gap semiconductor (3.37 eV), like GaN and ZnS, with large exciton binding energy (60 meV) that exceeds the room temperature thermal energy (25 meV). Owing to this favorable property, it is characterized by highly efficient exciton emission at room temperature [1]. ZnO is transparent to visible light and operates in the UV wavelength. ZnO band gap can be engineered by impurities (doping by Mg increases the band gap, leading to deep UV, whereas dopants such as Cd decrease the band gap, leading to blue emission) [2]. Several research groups attempted to improve its epitaxial growth technologies and nanostructure synthesis in order to produce high quality ZnO materials for optoelectronic and electronic devices [3]. Due to its unique electrical and optical properties, ZnO can have many potential applications, including gas sensors, UV photo-detectors, high transmittance conductive oxide coating and bulk acoustic wave resonators. In addition, ZnO has significant industrial applications, such as in production of plastics, rubber, paints, batteries and cosmetics [4]. Furthermore, ZnO can be used in space and nuclear applications due to its high radiation resistance [4, 5]. It can be grown on different substrates, such as glass, silicon, and sapphire. ZnO plays a significant role in a wide range of biomedical, bio-safe and biocompatible applications.

1.1 ZnO Nanostructure

Nanotechnology is the science or technology that studies materials at the nanometer scale. The national technology initiative defined nano-sized particles as 1 nm to 100 nm in dimension [6]. As nanomaterials possess an intermediate structure, between molecule and bulk, they have unique physical and chemical properties, distinct from those of their bulk form [6]. The physical properties of nanomaterials depend on the types of compounds, impurities, defects and the bond types. However, due to the quantum confinement within a nanostructure, the charges and electrons have confined behavior. This confinement causes discrete energy levels and wider band gap compared to the bulk semiconductors [7]. The quantum confinement of nanostructure is much larger than that of the bulk, leading to structural distortions that arise because of the presence of a greater number of dangling bonds, which make nanostructures sensitive to chemical bonds and electrical environment changes [7]. In terms of structural properties, interatomic spacing in nanostructure and the bulk is different because the surface area and surface energy increase with decreasing particle size [8]. The change in structure as a function of particle size also affects the optical properties of the material. For example, the band gap increases as the particle size decreases (due to larger quantum confinement effect compared to a larger particle), leading to the inter-band transition shifting to a higher frequency. Owing to these characteristics, the emissions from semiconductor nanostructures can be adjusted by changing the nanostructure size [9].

Recently, ZnO nanostructures (such as nanowires, nanorods, nanotubes and nanoplates) have become the subject of extensive research due to their larger surface area to volume ratio compared to its bulk form. ZnO nanostructures are attractive candidates as components in photonic device applications, such as ultraviolet photodetectors, sensors, field effect transistors, intermolecular p-n junction diodes, and Schottky diodes. Consequently, due to the many applications and devices incorporating ZnO and its invaluable properties, ZnO is used as research material in this work. It is envisaged that devices based on ZnO will be the most functional in the near future [10-13]. This research has been performed with the aim of obtaining nanostructures using the "bottom-up" approach, without using catalysts or metal seeding/nanoparticles. As one aspect of this research, nucleation and growth processes were investigated for ZnO nanostructures grown by pulsed laser deposition (PLD), focusing on their unique physical, structural and optical properties resulting from their low dimensionality.

1.2 Europium-doped ZnO

In 1896, French chemical scientist Eugène-Antole Demarçay discovered Europium (Eu), a rare earth element[8]. Eu is the most reactive rare earth element, with an atomic number of 63 and atomic weight of 151.964. Because of its red radiation, Eu is widely used in television screens and computers. This red rare earth emission is very sharp [14] and, unlike semiconductor band edge emission, it does not have strong temperature dependence because its emission is due to intra-4f transitions that are deeply buried by 6s and 5d levels. As a result, it produces a pure red line, which is ideal for devices operating at room temperatures [8].

11

1.3 Thesis objective and contents

The main objectives of this book are synthesizing ZnO nanorods using PLD without any catalyst and investigating the effects of PLD process conditions on the morphology and optical properties of ZnO nanorods. In addition, the structural and the luminescence properties of Eu-doped ZnO nanostructure are studied, while also attempting to understand the critical effects of the PLD growth conditions on these properties.

This book is comprised of six chapters. The introduction to the research reported in this study is provided in Chapter 1, while Chapter 2 presents a brief literature review, focusing on extant studies on the structural and optical properties of ZnO materials. In Chapter 3, the physical mechanism of ZnO growth, including nanostructure growth, is introduced, followed by a brief description of the nucleation theory. Chapter 4 presents the experimental techniques used in this work. The work performed on ZnO nanorods that includes the effect of the PLD growth conditions on obtaining high quality vertical ZnO nanorods and their structural and luminescence properties are presented in Chapter 5. Finally, conclusions are given in Chapter 6.

Chapter 2
ZnO properties

In this chapter, a brief summary of the structure, as well as the optical and electric properties, of ZnO will be reported.

2.1 ZnO crystal structure

ZnO has a substantial ionic character, as do other II–a VI semiconductor compounds in which iconicity resides at the borderline between covalent and ionic semiconductors [10]. In the mixed ionic-covalent bonds, the shared electrons are located closer to the larger electronegative atom (anion) [15]. ZnO can be crystallized in three forms, namely as wurtzite (hexagonal), Zincblende (cubic) and rocksalt. The ZnO unit cells of these three structures are shown in Figure 2.1 [10].

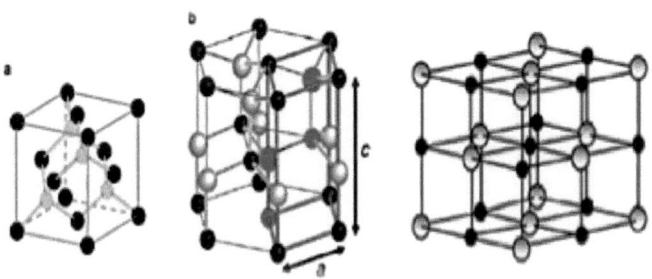

Figure 2.1. Units cells of three ZnO crystalline forms: (a) Cubic Zincblende, (b) hexagonal wurtzite [10], and (c) cubic rock salt structure [5].

Wurtzite ZnO: In this study, only wurtzite ZnO is of interest, as it is the most stable structure. The hexagonal unit cell is characterized by two lattice constants (a, c) (shown in figure 2.1), with the a/c ratio = 1.6333. The a-axis corresponds to the horizontal crystal direction, as shown in figure 2.1, whereas the c-axis direction is referred to as [0001] direction, perpendicular to the (1000) plane [10, 16]. Wurtzite ZnO structure has a space group $P6_3mc$, comprised of two interpreting hexagonal closely packed (hcp) sub-lattices. In each sub-lattice, one type of atom is displaced with respect to the others along three-fold c-axis symmetry [10, 16], as shown in Figure 2.1 (b). In wurtzite ZnO, three surfaces and plane (0001), ($11\bar{2}0$) and ($1\bar{1}00$) are of particular importance. Their corresponding directions are [0001], [$11\bar{2}0$] and [$1\bar{1}00$], respectively. The {0001} is a basal plane, and is typically used in epitaxial

13

growth [10]. These main directions and planes are shown in Figure 2.2 [13].

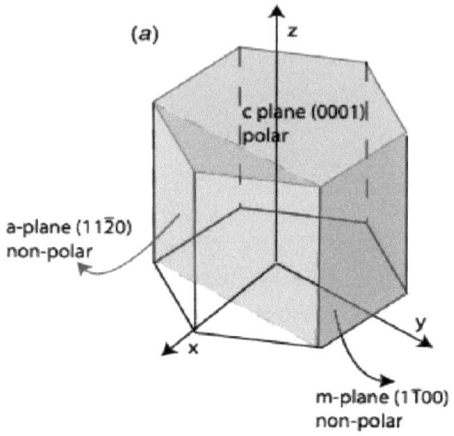

Figure 2.2. The most important planes in the ZnO crystal structure [13].

Many factors can play an important role in determining the lattice parameters of semiconductor materials [10], including (a) external and internal strain induced by substrate or impurities, (b) concentration of free electrons acting via deformation potential of a conduction band, and (c) concentrations of impurities and defects [10]. In wurtzite ZnO, the value of a-lattice constant varies from 3.2475 to 3.25 Å, while that of c-lattice constant varies from 5.2042 to 5.21 Å correspondingly the c/a ratio varies from 1.593 to 1.6035 [10].

Zincblende ZnO: As the zincblende ZnO structure is metastable, it can be stable only in the heteroepitaxial growth [10]; when the material of substrate is different from that of the film [17]. According to the Hermann-mauguin notation, the symmetry structure of zincblende has $F\overline{4}3m$ space group [10]. This structure is composed of two interpenetrating face-centered cubic (FCC) sublattices shifted along the body diagonal, as shown in Figure 2.1 (a). Each of four oxygen atoms in the unit cell is surrounded by four Zn atoms. This tetrahedral coordination formed by 4 nearest neighbors and 12 next neighbors is present in both wurtzite and Zinc blende structure. Consequently, the bond distance is identical in both crystalline forms [10].

Rocksalt ZnO: At high pressure (10 GPa), the structure of wurtzite ZnO (like other II-VI semiconductors) can be transformed to that of rocksalt (NaCl structure) [10]. As ZnO has space group Fm3m in this case, the rocksalt structure cannot be stabilized by the epitaxial growth [10].

2.2 Band gap structure

Figure 2.3 shows the band structure of wurtzite ZnO. ZnO is considered to possess a direct band gap because the top of the valence band (VB) and the bottom of the conduction band (CB) occur at the same momentum value (K= Γ), as shown in Figure 2.3 [18]. Moreover, it has a wide energy gap, E_g = 3.44 eV at low temperatures (10K), with slightly lower value of 3.37 eV at room temperature [19]. Decreasing the temperature leads to contraction of the crystal lattice that causes more strength to interatomic bonds and results in a wider band gap [20].

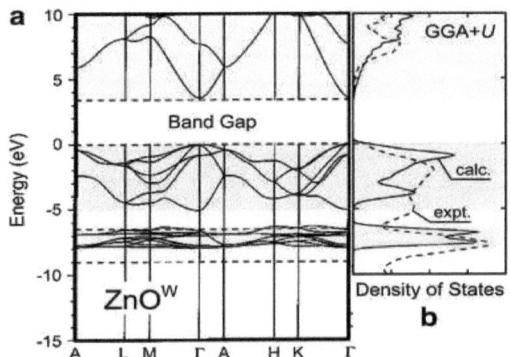

Figure 2.3. The calculated band structure of wurtzite ZnO using density functional theory with generalized gradient approximation (GGA+U) [2].

2.3 Optical Properties

Photoluminescence (PL)

Luminescence is a process that differs from reflection, refraction and scattering of the light. It is a radiative emission process that certain materials are capable of generating. PL occurs when the substance absorbs the excitation light and the energy of the light is transferred to an electron in the ground state (VB in semiconductors). Owing to the greater energy it possesses, this electron will move into an excited state (CB or impurity/defect level in semiconductors), followed by an electron-hole recombination that releases another photon, as shown in Figure 2.4 [21].

PL is an efficient process to study the optical properties of semiconductors [22]. PL technique assists in understanding several properties, including recombination mechanisms, determining the interference and impurity level of the semiconductors, identifying transitions near the band conduction and valence band edge, estimating the material quality [23], and classifying defects and impurity bands in the semiconductor band gap [24, 25].

2.3.1 PL Spectroscopy of Semiconductor Materials

Semiconductor emission can be described as band edge emission and defect-related emission. PL of semiconductor materials is categorized as either intrinsic PL or extrinsic PL, as shown in Figure 2.4.

Intrinsic optical properties of ZnO semiconductor are related to band-to-band transitions and excitons that occur between the electrons in the conduction band and the holes in the valence band. On the other hand, the extrinsic properties are associated with dopants or defects, which generate discrete electronic states in the band gap [22].

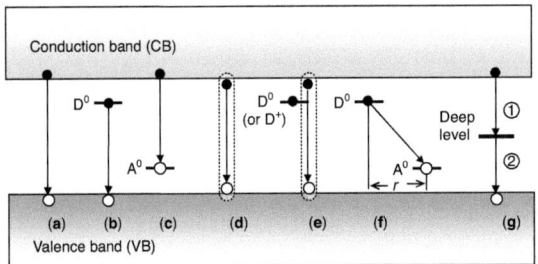

Figure 2.4. A diagram of radiative recombination process in a semiconductor, in (a) band-to-band recombination, (b) neutral donor D^0 to VB transition, (c) CB to neutral acceptor A^0 transition, (d) radiative recombination of FE, (e) radiative recombination of BE bound to D^0, (f) DAP recombination with separation r and (g) deep level defect luminescence [26].

Band edge emission occurs by the radiative band-to-band transition in the semiconductor at relatively high temperatures, as shown in Figure 2.4 (a). An electron is excited in the CB and a hole is excited in the VB by the absorption of the photon to give rise to radiative recombination. At low temperatures, due to coulomb attractive force, excitons (electron-hole pairs) are formed, rather than band-to-band transition, as shown in Figure 2.4 (d). In extrinsic semiconductors, impurities are incorporated as either acceptors or donors, whereby PL occurs via defects states (defect emission) [27]. These mechanisms will be described in detail in the following sections.

a) Band edge emission

Band edge emission contains exciton emission (including bound exciton and free exciton). Exciton emission can be either free (FE) or bound exciton (BE). The latter type refers to that associated with both neutral or charged donors and acceptors [29]. A Typical PL spectrum of ZnO is shown in Figure 2.6.

I. Bound exciton (Frenkel):

When a semiconductor absorbs photon energy exceeding the band gap, a pair consisting of an electron (in the CB) and a hole (in the VB) is created by coulomb interaction and forms a new particle called exciton. When an exciton is localized near an acceptor/donor it is called exciton bound to acceptor/donor, as shown in Figure 2.5 [24].

II. Free exciton (Mott-Wannier):

When the radius of the exciton is large it is referred to as free exciton, in the sense that it is not localized near any atoms, as shown in Figure 2.5 [21, 28].

III. Free-to-bound transition:

A free electron from the conduction band can be bound to an acceptor, as shown in Figure 2.4 (c). Similarly, a free hole from the valence band can be bound to a donor (Figure 2.4 (b)) [24]. Donor-acceptor pair recombination (DAP) is shown in Figure 2.4 (e).

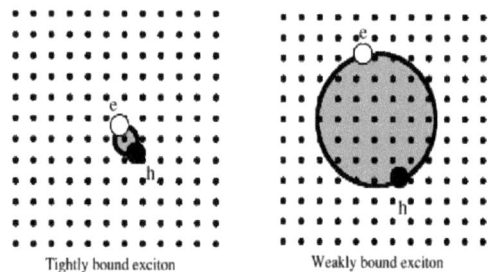

Figure 2.5. Graphical representation of (a) bound exciton and (b) free exciton [21].

b) Defect bands

Deep level luminescence is a transition that occurs when some defect impurities create deep levels inside the band gap. Figure 2.4 (g) illustrates the recombination process of the deep-level luminescence [27].

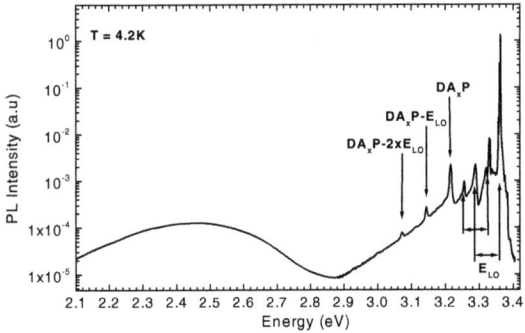

Figure 2.6. PL spectra of ZnO excitonic, donor acceptor pair (DAP) and deep level emission [30].

2.3.2 UV ZnO emission (band edge emission)

The ZnO band edge emission is UV centered at 368 nm [31], resulting from the wide ZnO band gap. In the band edge ZnO spectra at low temperatures (4-10 K), the bound exciton emission is most apparent, whereas free exciton predominates at room temperature [32]. In addition, PL band edge emission contains donor bound exciton and excitons bound to neutral acceptors a donor-acceptor pair transition as shown in Figure 2.6 [32]. Table 2.1 illustrates the energy levels of different excitons.

Table 2.1. Emission energy of ZnO peaks at low temperature [22, 32].

Emission type	Emission energy (eV)
Donor bound exciton	≈ 3.3628
Acceptor bound exciton	≈ 3.3564
Donor– acceptor pair transition	≈ 3.218 - 3.223
Free exciton	≈ 3.377 [5]
Bound exciton	≈ 3.3484 - 3.3693

19

2.3.3 Defect Emission

Figure 2.7 shows the defect energy levels within the ZnO band gap that can be influenced by both optical absorption and emission processes [33]. Several peaks in the visible spectral region can be identified in the PL spectra of ZnO, which correspond to the defect states in the band gap (405, 420, 446, 485, 510, 544, 583 and 640 nm) [32]. Figure 2.7 shows a graphical representation of the defect energy states (levels) in the ZnO band structure.

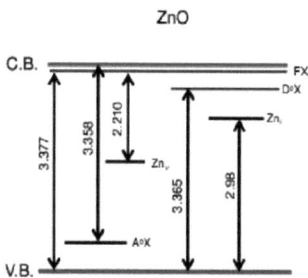

Figure 2.7. Defect energy (states) levels in ZnO band gap, where the peak at 3.377 eV is attributed to bound exciton, with the remaining peaks corresponding to different defects in ZnO [32].

Green band emission centered at ≈535 nm [34] is the most common band emission and is attributed to oxygen vacancies (Vo) and zinc interstitial defects (Zn_i) located at the surface of the ZnO material [32, 35]. Similarly, orange-red emissions centered at ≈540 nm [32] are attributed to oxygen interstitials (O_i) and zinc vacancies (V_{Zn}) [32] [36]. Finally, violet defect emission is attributed to the presence of shallow (Zn_i) defects centered at ≈435 nm [29, 32].

2.4 Doping in ZnO

Doping refers to the process of adding impurities (foreign atoms) into the host material. The aim of impurity doping is to optimize the physical properties of the material [7], and it can thus be an effective way of controlling the electrical conductivity, optical luminescence and magnetic properties [37]. For example, in terms of its conductivity, stoichiometric ZnO is classified as insulator. Thus, effective utilization of native defects (such as Zn_i, V_o [38], or addition of Al dopants, can increase the ZnO conductivity by about 10 orders of magnitude [19].

2.4.1 n-type doping

Zinc oxide is a naturally n-type semiconductor. The intrinsic defects that can be responsible for the n-type conductivity, formed during deposition, include Zn_i and V_O [33]. In practice, n-type ZnO semiconductors are easily obtained by, for example, doping with various impurity materials, including group-III elements (Al, Ga, In), and traditional metals (Pb, Mn, Fe, Ni, Co). When combined with the ZnO lattice, these dopants form shallow donor levels [33].

2.4.2 p-type doping

Unlike n-type, p-type ZnO is difficult to produce. These difficulties arise due to many factors, including the tendency toward n-type conductivity, low-energy native defects compensating the attempted p-type dopant (such as Zn_i or V_O) and some of the centers of compensation [10].

Many elements can be used in doping ZnO, where they will act as acceptors. Group-I elements (Li, Na and K), copper (Cu) silver (Ag), Zn vacancies and group-V elements (such as N, P and As) can create deep acceptors [19].

2.4.3 Doping with Rare Earth Elements

Since the 1970s, rare earth (RE) doped ZnO has been widely utilized for different photoelectronic applications, which led to significant improvements in semiconductor technology. Rare earth elements (RE) are fifteen metallic elements of the lanthanide series, as shown in Figure 2.8 [39]. Their name refers to the difficulty of obtaining these materials in commercially viable concentrations. REs are classified into two subgroups: light rare earth elements (LRE) comprise the first five elements (La- Sm) and the heavy rare earth elements (HRE) comprise the elements (Eu - Yb), as well as yttrium. Despite its low atomic weight, yttrium is classified as HRE, based on its properties [40]. In the periodic system depicted in Figure 2.8, positions of these elements (lanthanide series) are highlighted.

In this study, only Eu-doped ZnO grown on a-sapphire is considered. Figure 2.9 shows the red emission of Eu^{3+} in Eu-doped ZnO due to intra 4f transitions (from $^5D_0$ to ^7F_n levels) [41] .

Figure 2.8. Position of the rare earth elements in the periodic table[36].

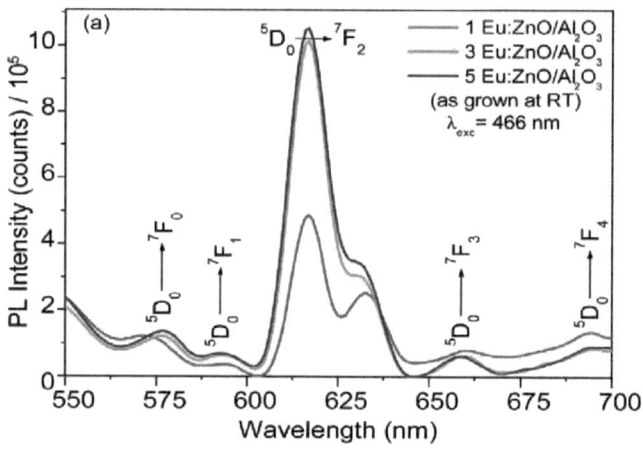

Figure 2.9. Photoluminescence spectra of Eu-doped ZnO grown on sapphire [42].

Eu-doped ZnO thin films present Eu red emission, demonstrating energy transfer from the host to the dopant ions. Moreover, previous studies have shown that Eu-doped ZnO could gain high luminous efficiency, indicating that Eu-doped ZnO is a promising material for lighting and flat panel display applications.

In general, electrons can be excited by two methods, either via direct electron excitation, or by indirect excitation through carrier-mediated energy that is transferred from the semiconductor host to the electrons [38].

Direct excitation: By using an optical pumping source, such as photoluminescence (PL) excitation, the electrons will excite directly [38].

Indirect excitation: If the photon energy is greater than the band gap, the host transfers the energy via the recombination of an electron-hole pair at the carrier trap. In this case, coulomb force attracts the free electron or hole to the trap, as a positive or a negative charge, to produce a bound exciton [38], as shown in Figure 2.10.

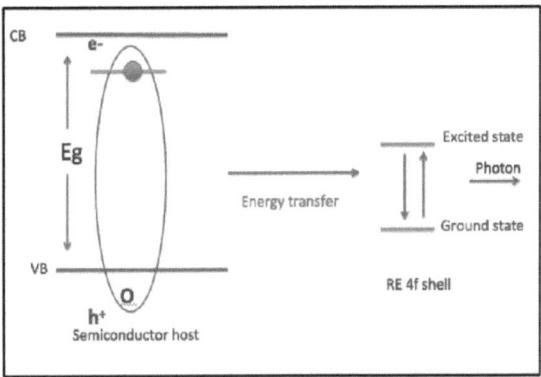

Figure 2.10. Shows indirect energy transfer from a semiconductor host to the electron ions

Chapter 3
ZnO Growths

This chapter presents ZnO deposition technique based on pulse laser deposition (PLD) method and its growth mechanism.

3.1 Introduction

ZnO can be grown by several techniques on different substrates, such as glass, sapphire (Al_2O_3) and diamond, including lattice-matched substrates, such as *a*-sapphire and (0001) $ScAlMgO_4$, which have 0.8 and 0.09% lattice mismatch with (0001) ZnO, respectively. High quality ZnO nanowires and nanorods are mainly fabricated by various chemical and physical vapor deposition techniques [10], such as metal-organic chemical vapor deposition (MOCVD), thermal evaporation, metal-organic vapor phase epitaxy, PLD, molecular beam epitaxy (MBE) and sputtering [40].

In general, there are two types of epitaxial thin films namely, homo-epitaxial and heteroepitaxial [43]. Homoepitaxial growth is achieved when the deposited material is the same as that of the substrate, such as n-Si on Si substrate. This process is used to produce high quality films, especially if the material will be doped by impurities due to the absence of the line defects (extended defects or dislocations) [44]. Heteroepitaxial growth is the process of depositing a material onto a substrate that has atoms of different types [43]. In heteroepitaxial growth, as lattice mismatch between the film and substrate is viable, it can produce low quality materials with line defects [43,44].

3.2 Substrates

In ZnO, heteroepitaxial growth can be achieved using different lattice-mismatched (CaF_2, Si, GaAs, *c*-Al_2O_3) and lattice-matched substrates ($ScAlMgO_2$, *a*-Al_2O_3). Closely lattice-matched substrates are preferred when growing high quality ZnO. For example, *a*-sapphire substrate is preferable for ZnO heteroepitaxial growth. The strain and dislocation density can be reduced significantly by growing (0001)-oriented ZnO on the ($11\bar{2}0$) *a*- Al_2O_3 [10]. These benefits stem from the lattice match between *a*-sapphire and (0001) ZnO, with only 0.08% mismatch [42]. In Table 3.1, the structural properties of different substrate types that can be used for ZnO deposition are presented [10].

In the present study, *a*-sapphire substrate was primarily used. However, for comparison, in some deposition conditions, ZnO and *c*-sapphire substrates were also employed.

Table 3.1. Lattice parameters of different substrates for ZnO [10].

Substrate Material	Crystal structure	Lattice parameter (nm)	Lattice mismatch (%)	Space group
ZnO	Hexagonal	a =0.3249 c = 0.5206	0	$P6_3mc$
$c\text{-}Al_2O_3$ $a\text{-}Al_2O_3$	Hexagonal	a = 0.4754 c = 1.299	~32 0.08	$R\bar{3}c$
GaAs	Cubic/Zincblende	a = 0.5653	–	$F\bar{4}3m$
GaN	Hexagonal	a = 0.3189 c = 0.5185	-1.9	$P6_3mc$
MgO	Cubic/rocksalt	a = 0.4216	–	Fm3m

3.2.1 Sapphire Substrate

Sapphire (Al_2O_3) is a ceramic material that is crystallized in both rhombohedral and hexagonal structures, with a unit cell volume of 84.93 Å^3 and 254.79 Å^3, respectively [10]. Figure 3.1 shows the difference between two sapphire Al_2O_3 lattice structures. The focus of the present study is on hexagonal Al_2O_3, as it is a suitable substrate for (hexagonal) wurtzite ZnO. In particular, a-Al_2O_3 is of interest, due to its small mismatch with ZnO grown along the c-axis [10]. Hexagonal sapphire has 30 ions in the total form of the lattice, with 12 Al^{3+} and 18 O^{2-} ions [10].

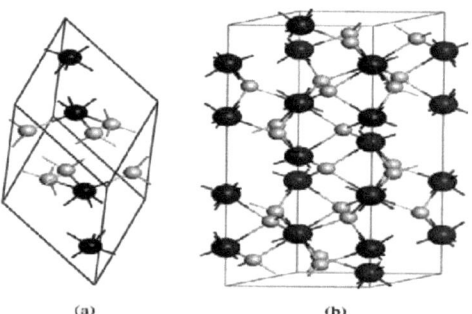

(a)   (b)

Figure 3.1. Rhombohedral (a) and hexagonal (b) sapphire unit cell where the gray atoms are for Oxygen and the dark atoms are for Al [10].

While sapphire substrates can be cut in several orientations, the most important ones are c, r, a, and m [45], as shown in Figure 3.2. The c-plane cut is achieved along [0001] direction, which is the basal plane for hexagonal structure, as shown in Figure 3.2, where the a-plane cut is along [11-20] direction [46].

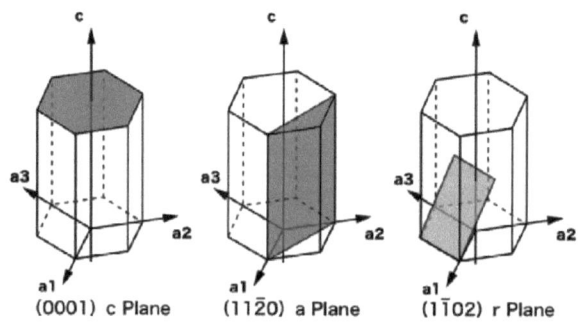

(0001) c Plane ($11\bar{2}0$) a Plane ($1\bar{1}02$) r Plane

Figure 3.2. Orientation of sapphire planes [47].

3.3   Growth process

In this study, PLD is used to deposit ZnO nanostructure and thin film on *a*-sapphire Neocerra PLD system (Neocerra, USA) employing KrF excimer laser (λ=248 nm) in the core labs facility at KAUST. As the PLD technique will be described in detail in Chapter 4, only the general growth mechanism and the physics background are presented here. The deposition process occurs in the following sequential steps: (i) the transferring of the ionized species from the target and adsorbed on the substrate surface. (ii) Diffusion process of species occurs before their incorporation to the film. (iii) The diffused species attract each other and are attracted by the substrate surface to form bonds of the created materials (films or nanostructures). (iv) Nucleation takes place. As the initial aggregation of the material continues, it increases the film's thickness or grows nanostructured crystals. In order to create the desired structure and properties, controlling many kinetic processes, such as surface diffusion, surface energy and nucleation are important. Each of the aforementioned processes is described briefly in the following sections.

3. 3.1   Surface diffusion

Surface diffusion is defined as the mobility of the species of grown materials from the surface to the whole material. It is important in determining the structure and the crystallinity of grown materials for both film and nanostructure depositions because, via surface diffusion, the adsorbed species can move to the most active sites and create the material structure [48].

3.3.2   Surface energy (γ)

Surface energy determines if the materials will be formed as films or nanostructures. In thermodynamics, surface energy is defined as the energy required for creating a unit area in the new surface. The stability of any system or material is defined by the state with the lowest Gibbs free energy, the minimum energy required for the system to complete its process spontaneously. As a result, materials tend to minimize the total surface energy [48]. The surface energy can be defined as a function of Gibbs energy [43]:

$$\gamma = (\partial G / \partial A) \text{ n, T, P} \qquad (3.1),$$

where G is the Gibbs energy, A is the area, n is number of atoms T is the temperature, and P is the pressure. Surface energy can be also defined as the energy required for the surface atoms to occupy to their original positions (ideal surface positions) when their surface energy is at the minimum. It can thus be defined as [43]:

$$\gamma = \frac{1}{2} N_b \, \varepsilon \, P_a \qquad (3.2),$$

where N_b is the number of broken bonds, ε is the bond strength, and P_a is the surface atomic density (the number of atoms per unit area on a new surface).

3.3.3 Nucleation

During deposition on a substrate, nucleation behavior is strongly influenced by the surface energy of the substrate γ, which determines the final structure of the materials (e.g., one-dimensional (1-D), two-dimensional (2-D) or three-dimensional (3-D) growth). Of particular interest here are the surface energy of the substrate (γ_s), the surface energy of the grown crystal (γ_c) and the interference energy between the surface and the substrate (γ_i). These three values depend on the material's crystallographic orientation, atomic reconstructure (the atoms in the surface have a different structure than those in the bulk), chemical composition and atomic scale roughness, as well as the growth conditions (e.g., the energy of the deposited ionized species and the growth temperature). Three types of nucleation processes can be recognized during the growth process [48]:

(i) 2-D Smooth growth (Frank van der Merwe process/method): In this process, the deposited species cover or wet the substrate completely, which leads to 2-D or film deposition, as shown in Figure 3.3 (a). This process occurs when the substrate surface energy is greater than the sum of the free surface film and the film interface energy $\gamma_i + \gamma_c < \gamma_s$. For this kind of growth, strong bonding between the film and the substrate is required, as the aim is to reduce γ_i [48].

(ii) 3-D nucleation (the Volmer-Weber process/methods): In this process, while the film does not cover or wet the substrate, it forms 3-D islands, as shown in Figure 3.3 (b). As no bonding takes place, in this case, the total surface energy is given by $\gamma_i = \gamma_c + \gamma_s$ [48].

(iii) Island formation (Stranski-Krastanov growth): In this case, the island formation commences before one or few monolayers of deposit material cover the substrate surface, as shown in Figure 3.3 (c) This kind of growth is induced by stress due to the lattice mismatch between the substrate and the deposited species [48].

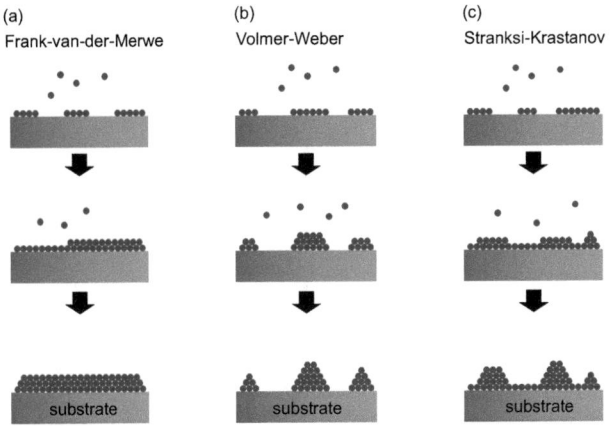

Figure 3.3. Three different growth mechanisms: (a) Franc-van-der Merwe or layer-by-layer growth, (b) Volmer-Weber or island growth, and (c) Stranksi-Krastannov [48].

The growth of film or nanorod structures can be achieved by manipulating one of the surface energies (γ_i or γ_c). For example, if γ_i decreases, the film substrate bonding increases, which leads to wet deposition and a formation of a film. This bond is affected by the chemical reactivity between the film and the substrate atoms, as well as between the atoms of the deposited materials [48].

The relationship between the area and the surface energy of the deposited material is important because the surface energy of the grown crystal is different depending on the structure [43]. Therefore, in PLD deposition, by controlling the growth kinetics, it is possible to change the ZnO structure. In this study, the deposition parameters such as temperature, pressures, and laser energy and the target-substrate distance can be changed with the aim of controlling the diffusion, growth type and the final structure of ZnO. These processes will be described in the subsequent chapters.

3.4 ZnO Nanostructures

ZnO nanostructures can be divided into four groups: (i) 3-D structures, such as a bulk structures (nucleation); (ii) 2-D structures, such as thin films and monolayers (these structures have two dimensions that exceed the nanometric size range [49]); (iii) 1-D structures (with dimensions within the nanometric size range), such as nanorods, nanoneedles, nanotubes and nanowires; and (iv) 0-D (or dimensionless) structures, also known as atom clusters [15]. Various types of ZnO nano- and microstructures reported in the literature are shown in Figure 3.4 [13].

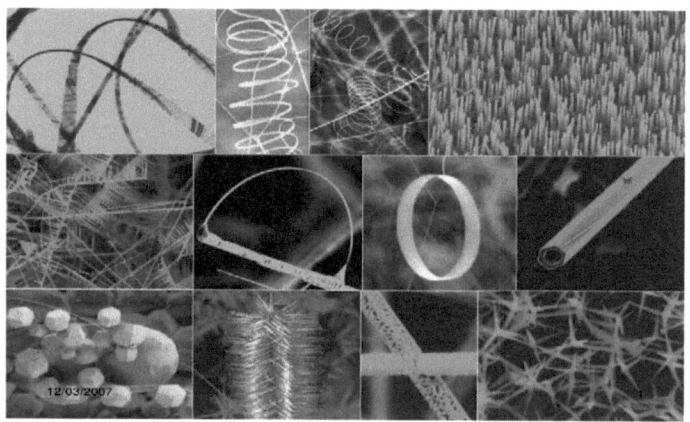

Figure 3.4: Different ZnO nanostructures and microstructures [50].

3.4.1 One-dimensional nanostructures

In this study, ZnO 1-D nanostructure is presented. Thus, the methods that are used for synthesizing this 1-D structure will be described in this section.

I.   Spontaneous growth:

This method is employed in this present work. It is known that its success depends on the reduction of Gibbs free energy or chemical potential. Gibbs free energy is realized by phase transformation, chemical reaction or release of stress. For a given material and particular growth conditions, in spontaneous growth, defects and impurities affect the final morphology of the material. There are three techniques by which spontaneous growth can be achieved, which are differentiated by the spontaneous growth mechanism: evaporation–condensation, vapor-liquid-solid (VLS or SLS) growth, and stress-induced recrystallization [42].

II.   Template-based synthesis:

This is a general method used to fabricate nanorods, nanowires and nanotubes of polymers, metals, semiconductors and oxides. Templates with various nanosized channels are used to grow nanotubules and nanorods. Template-based synthesis incorporates chemical deposition methods, such as electroplating and electrophoretic deposition, colloid dispersion, melt, or solution filling conversion with chemical reaction [51].

I.   Electro-spinning:

This approach is also known as electrostatic fiber processing, as it was originally used for generating ultrathin polymer fibers. It uses electrical forces to produce polymer fibers of nanometer-scale diameters. This phenomenon occurs when the electrical forces at the surface of a polymer solution or melt overcome the surface tension and cause an electrically charged jet to be ejected [15].

II.   Lithography

This is a top-to-bottom post-growth method that can be used to fabricate nanostructures after growing thin films. Its main application is in fabrication of nanowires with diameters ranging from 10 nm to 100 nm [52].

Chapter 4

Experimental Techniques

In this chapter, the PLD technique used in this work for growing ZnO nanostructures is described. The background and the working principle of several characterization techniques employed as a part of this study are also presented, namely scanning electron microscopy (SEM), x-ray diffraction (XRD) and photoluminescence (PL).

4. 1   Pulsed laser deposition (PLD)

The PLD technique was first used in 1980 as a film deposition technique, based on physical evaporation [53]. The PLD process is based on a plasma plume of highly energetic ionized species created (atoms and ions evaporated from the target) by interacting with a high-energy laser beam [54]. This interaction takes place inside a vacuum chamber. Figure 4.1 shows a standard PLD setup, which includes a pulsed laser beam focused onto a material target by using a group of optical lenses. Material target is sintered powder of different materials used for deposition. Light-matter interaction occurs between the focused pulsed laser beam and the target's surface to generate ionized species (plasma plume) [54].

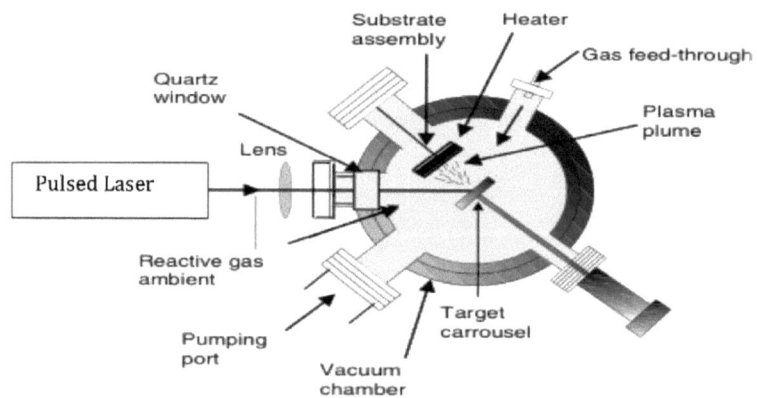

Figure 4.1. Pulsed laser deposition setup [5] .

4.1.1 PLD deposition mechanism

The full PLD process is realized in the following sequence:

1. The interaction between the laser and the surface target. In this stage, the laser is focused onto the target surface under specific conditions (substrate temperature, laser energy density, pulse duration, pressure, the distance between the substrate and the target) that are determined before starting the deposition process. Species (atoms and ions) are dissociated from the target and are ablated out by stoichiometry. The rate of the ablation depends on the pulsed laser density [53].

2. Kinetic motion of the ablated materials: In this step, the ablated species tend to move toward the substrate according to the gas dynamics law [53]. The substrate temperature and the laser spot size determine the uniformity of the film growth, while the distance between the target and the substrate controls the spreading of the ablated species [53].

3. Decomposition of the ablation species onto the substrate: The high-energy ablated species hit the substrate surface, whereby they sputter some of the surface atoms to create collision region. This region acts as a source for condensation of particles. The creation of the film starts when the rate of condensation is higher than the arrival rate of particles supplied by sputtering [53].

4. Nucleation followed by epitaxial growth on the substrate takes place as the final step [53].

a) Advantages of PLD

The following beneficial characteristics make PLD favorable relative to other deposition/growth techniques:

- PLD is a clean technique and capable of producing multilayer films at low cost compared to other common techniques such as MOCVD and MBE [55].
- Controlled stoichiometric transferring of material from the target to the substrate; energetic species that is generated in the plasma background gas; and the ability to create multi-cation films (such as ZnMgO or Eu ZnO) by using single stoichiometric target [56].
- This approach is versatile, because several types of materials can be deposited using PLD, such as metals, polymers, semiconductors, and ceramics [55].
- Controlling the growth conditions and laser energy can be easily optimized in the deposition process. In addition, each component in the plume has a similar deposition rate and energy [55].

- PLD can ease manipulation of target, substrate rotation and multi target deposition.

b) Disadvantages of PLD

The aforementioned advantages notwithstanding, PLD also has the following drawbacks:

- Thin film growth by PLD is size-limited, due to the limited laser spot size and the small cross-section of the ablation plume (around one cm^2) [55].
- The thickness is difficult to monitor and control. The plume of the ablated material is highly forward directed, which causes poor coverage [55].
- Splashing or creating droplets or big particles of the target material on the substrate surface [55].

4.2   Scanning Electron Microscope (SEM)

In this study, SEM is employed in the investigations of the surface morphology. When using an electron microscope, microstructure of the materials at higher magnification and resolution can be achieved than that by conventional optical light microscope, as the wavelength of electrons is much smaller than that of visible light [57].

4.2.1   Background
a) Resolution of the Electron microscope

The microscope resolution ΔR is defined as the shortest distance between two points inside the specimen that are perceived in the image [58]. Higher resolution can be achieved by shorter wavelengths, using materials with a greater refractive index, and placing samples closer to the lens, as described by the expression below [58]:

$$\Delta R = \frac{0.61\lambda}{r_m \sin \theta} \qquad (4.1)$$

Where λ is the wavelength of the radiation used, r_m is the refractive index of the medium and $r_m \sin \theta$ refers to numerical aperture. Numerical aperture is equal to the sin of the half angle θ of the cone of rays coming to focus, multiplied by the reactive index [59]. For example, if the microscope light has the wavelength in the visible light range (i.e., 400-800 nm) and ΔR= 200 nm, the relationship between the wavelength and the electron energy that can be used to calculate the length can be expressed as [58]:

$$\lambda = \frac{h}{(2m_\circ eV)^{\frac{1}{2}}} = \frac{h}{[2m_\circ eV(1+\frac{eV}{2m_\circ c^2})]^{1/2}} \quad (4.5)$$

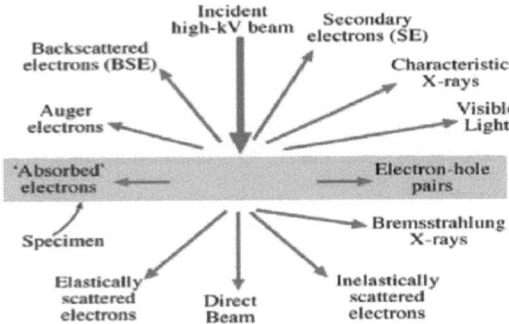

Figure 4.2.Various signals generated during interaction between the electron beam and the specimen [58].

b) Electron Specimen Interaction

Figure 4.2 shows various types of signals that can be generated by electron-matter interaction [58]. In this study, secondary electron (SE) and back-scattered electron (BSE) images are utilized, as the aim is to examine the surface morphology only.

When a high-energy (several keV) electron beam hits the specimen surface, the beam scattering occurs, generating BSE and SE, which are used as a signal source for surface imaging. Figure 4.3 illustrates the interaction volume of electrons and specimen atoms under the surface [57].

In Figure 4.3, the interaction zone according Monte Carlo simulation prediction is shown. It is pear-shaped and its volume depends on the electron beam voltage and material atomic number (Z) [60].

SE scans provide information on the area near the specimen surface (5-50 nm), while BSEs can be used to obtain detailed structure of the deeper level (about 50-300 nm). Figure 4.4 shows the difference between BSE and SE scans [58].

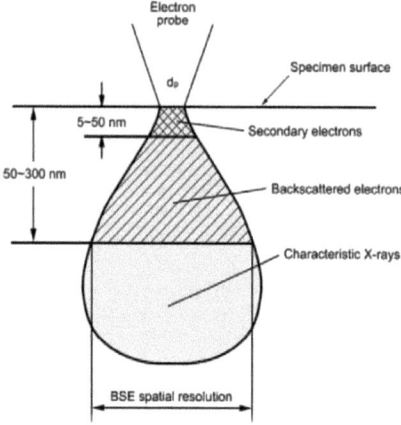

Figure 4.3. Interaction volume between the electrons and the specimen atoms under the surface [57].

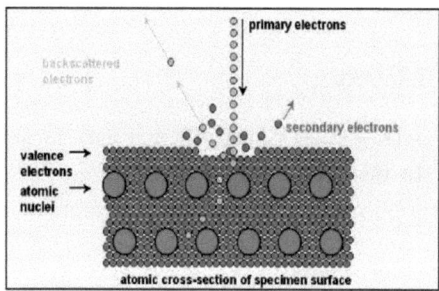

Figure 4.4. The difference between the excitation mechanism of SE and BSE [58].

4.2.2 SEM microscope setup

SEM consists of an electron gun, which is located in the top of the column. This gun generates electrons and accelerates them within the 1-30 keV energy range. Electromagnetic lenses and apertures are used to focus the electron beam and form a small spot size (1-100 nm) on the material surface [58]. High vacuum is required to avoid any scattering of the electron beam by the air. Figure 4.5 describes SEM components [58].

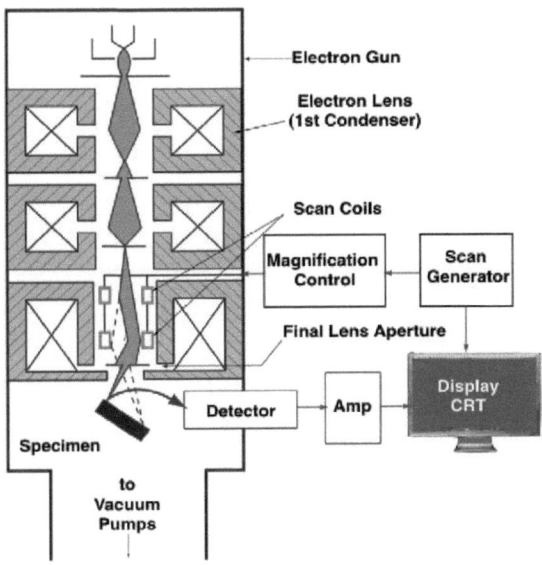

Figure 4.5. Scanning electron microscope components [57].

4.3 X-ray diffraction (XRD)

X-rays are electromagnetic waves with energies ranging from about 100 eV to 10 MeV. Their wavelengths range from about 10 to 10^{-3} nm [61]. X-rays are produced when high-speed electrons accelerated by a high-voltage field collide with a metal target [57].

X-ray diffraction is a technique used to determine the crystal structure of materials. It enables determination of the chemical compounds based on their crystalline structure [57].

The crystallographic planes in the material diffract the incident X-ray beams, as illustrated in Figure 4.6. In the example below, the two incoming beams (A and A') are deflected by two crystal planes. When Bragg's Law is satisfied, the deflected waves will be in phase, i.e., they will meet the following condition: $n\lambda = 2d \sin \theta$ [61].

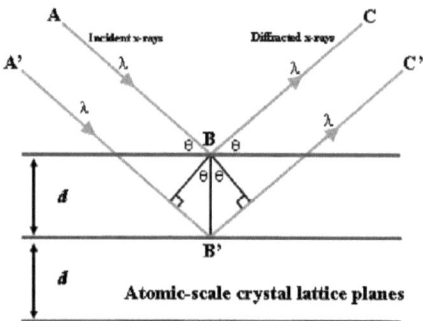

Figure 4.6. Graphical representation of the diffraction planes (Bragg's Law) [62].

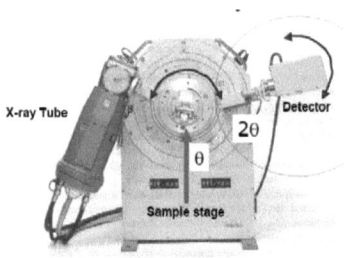

Figure 4.7. XRD diffraction setup [63].

By calculating the path differences between the two beams, the Bragg's law can be obtained, as the path difference depends on the incident angle (θ) and the spacing between the parallel crystal planes (d). The XRD diffraction setup consists of an X-ray source, a sample stage and a detector, as shown in Figure 4.7 [63].

XRD characterizations allow us to study information such as the quality of the materials, the crystal structure of the materials, the orientation of the crystals, the type of the crystal lattice, the lattice parameters, the chemical compositions of the materials and the strain in the materials [61].

4.4 Photoluminescence (PL)

4.4.1 Photoluminescence setup

Photoluminescence set up consists of the laser beam passing through the lenses, allowing it to focus onto the sample and excite the electrons in its material. The PL signal thus emitted from the sample passes through the lenses, which focus the emission onto a monochromator attached to a CCD detector, used to detect and record the PL spectra. Finally, the detector output is transferred to a computer, where the analysis is performed [24]. Schematic of the PL set-up is shown in Figure 4.8.

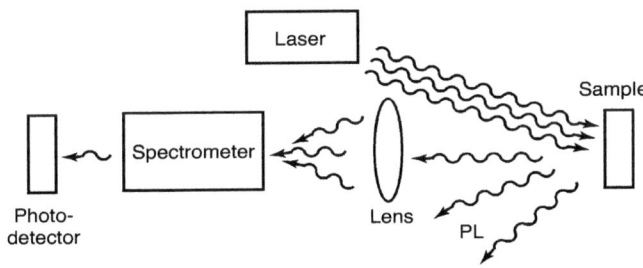

Figure 4.8. A standard PL setup [24].

Chapter 5

Results

This chapter presents the main results obtained in the present study. It commences with the description of the target preparation and the PLD conditions used for obtaining undoped ZnO and Eu-doped ZnO nanostructures. The structural (XRD and SEM) and luminescence properties of these materials will be presented before concluding the chapter with a discussion of the results obtained and the key findings.

5.1 Sample Preparation and Deposition

5.1.1 Target preparation

First, the Eu-doped ZnO and undoped ZnO PLD targets were prepared for use in the PLD depositions in our laboratory. This process consisted of several steps, namely measuring the powder quantity as per the stoichiometry (determining number of atoms in one Mole of material), powder mixing, pressing and sintering. More specifically, first, based on the ZnO_{1-x} $Eu_2O_{3\,x}$ formula, a predetermined amount of Eu_2O_3 (Sigma Aldrich, Europium (III) oxide, 99.999%) was added to a predetermined amount of ZnO (Sigma Aldrich, Zinc oxide, 99.999%, Trace metals basis) to prepare Eu-doped ZnO mixed powder. While these calculations depend on the atomic weights of the constituent cations, oxygen weight is ignored, because it is difficult to determine its percentage precisely when the target preparation process occurs in the air. The main objective here is to ensure high quality of target fabrication. ZnO and Eu-doped ZnO ceramic pellets were sintered at 1050°C for 24 hours to prepare the PLD targets. The density of the doped (0.2 wt% Eu_2O_3) doped ZnO target, is 5.18 g/cm^3, which gives a target diameter of 26.5 mm with a thickness of ~ 3 mm. The density of the Eu ZnO (0.15 wt% Eu) target is 5.11 g/cm^3, which produces a target diameter of 26 mm with a thickness of 2 mm. While the target quality is dependent on the characteristics of the homogeneous mixture of the constituent oxides, ZnO and Eu_2O_3 and its density must be close to the theoretical density of pure ZnO (5.61 g/cm^3). Empirical evidence suggests that a target with density 90% to 95% of that of pure ZnO is ideal for PLD ablation.

5.1.2 The target mixture calculation

In the present study, the commonly used ceramic target method is adopted. In this approach, the target size of 1 in × 1 in × 1 in and the minimum thickness of 3 mm are deemed desirable, along with a smooth surface and an immaterial shape. It should be noted, however, that the quality of the deposited thin film is influenced by the sintering conditions of the target [5]. In Tables 5.1 and 5.2 the calculation and results obtained for the target concentrations are presented. This process consisted of the following steps:

a) Calculating the number of Moles using the formula. $Zn_{1-x} Eu_2O_{3\,x}$.
b) Multiplying the result obtained in (a) with the molar mass of each compound in order to obtain Molecular × Wt.
c) Multiplying the results obtained in (a) with the value calculated in (b), yielding the oxide weight (in grams) and the total for all mixed compounds.
d) Dividing the weight (in grams) obtained above for each compound by the total weight, in order to derive the weight percentage (wt%).

Table 5.1. Calculation for Eu (0.2 at%) doped ZnO target:

Oxide	Mole per formula	Mol × Wt (G/Mol)	Total oxide weight (G)	Total weight (Wt. %)	For 10 g calculated (G)
ZnO	0.998	81.39	81.23	99.14	9.91
Eu_2O_3	0.002	351.92	0.7044	0.86	0.086
Total (mixture)	-	-	81.93	99.99	9.99

Table 5.2. Calculation for Eu (0.15 at%) doped ZnO target:

Oxide	Mole per formula	Mol × Wt G/Mol	Total oxide weight (G)	Total weight (Wt. %)	For 10 g calculated (G)
ZnO	0.9985	81.39	81.27	99.35	8.94
Eu_2O_3	0.0015	351.92	0.53	0.64	0.058
Total	-	-	81.79	99.99	8.99

5.1.3 PLD deposition process

In this work, a NEOCERA PLD chamber and the Lambda Physik KrF UV excimer laser (λ= 248 nm) were employed. In the experimental deposition, two Eu-doped ZnO targets with different nominal Eu concentrations (Eu_2O_3= 0, 0.15 wt% and 0.2 wt%) and Al_2O_3 on a-plane (11-20) substrates (from Surface Net GmbH, Germany) of 5 ×5 ×5.5 mm^3 in dimensions were used under similar PLD conditions. For each target, the effect of each growth parameter on the structural properties of the materials was studied, aiming to obtain nanostructures without a catalyst, thus 'improving on the currently available PLD methods. In order to achieve a comprehensive analysis, several deposition sets were prepared systematically. In each set, only one deposition parameter was changed, while the others remained fixed, thus facilitating the investigation of the effects of each deposition parameter on the material structure. The main objective here was to fully understand the conditions required for producing ZnO nanorods. The parameters that were investigated included growth temperature (T_g), laser energy (E_{PLD}), laser frequency (f_{PLD}), oxygen partial pressure (P_{O2}) and the distance (D) between the target and the substrate in the PLD chamber during the deposition. This systematic approach required about four months to complete and yielded approximately 40 samples. All the deposition sets are presented in Table 5.1.

Once each set (e.g., temperature dependence set) was deposited, the structure morphology was investigated via SEM, allowing the selection of the most optimal conditions for the production of the best nanostructure morphology in this particular set. This served as the input into the subsequent set (e.g., frequency dependence). By testing effects of each parameter sequentially, good quality Eu-doped ZnO nanorods structure was achieved in the final step. In these investigations, the results of the previous study of Eu-doped ZnO, conducted by the same research group, were used as the initial conditions. In Tables 5.3-5.7, the conditions used for obtaining each sample set in the present study are presented.

Having to complete this extensive work in only four months was a challenge, as many difficulties needed to be overcome. For example, obtaining the desired structure for Eu-doped ZnO nanorods was difficult, despite the fact that our research group has already successfully obtained vertically well-aligned ZnO nanorods using PLD. To our knowledge, this has never been accomplished before without using catalyst/nano seeding or modifying the PLD chamber. As many other important research projects were conducted in the PLD lab at the same time, it was not always available to carry out the depositions described above. The issue was further compounded by the fact that, at each deposition, the chamber and the PLD filters were cleaned and the laser energy measured, in order to ensure that the results were reproducible and the desired structure was obtained.

Table 5.3.The PLD conditions for set 1 (only the distance D was changed).

Sample set	Growth temperature T_g, (°C)	Pressure P_{O2}, (mTorr)	Distance D, (cm)	Laser energy E_{PLD}, (mJ)	Laser Frequency f_{PLD}, (Hz)	Number of pluses
1	650	35	10	300	10	30000
	-	-	12	-	-	-
	-	-	5	-	-	-

Table 5.4.The PLD conditions for set 2 (only the temperature Tg was changed).

Sample set	Growth temperature $(T_g, °C)$	Pressure $(P_{O2},$ mTorr)	Distance (D, cm)	Laser energy $(E_{PLD},$ mJ)	Laser Frequency (f_{PLD}, Hz)	Number of pluses
2	650	35	10	300	10	30000
	600	-	-	-	-	-
	700	-	-	-	-	-

Table 5.5.The PLD conditions for set 3 (only the laser frequency f_{PLD} was changed).

Sample set	Growth temperature $(T_g, °C)$	Pressure $(P_{O2},$ mTorr)	Distance (D, cm)	Laser energy $(E_{PLD},$ mJ)	Laser Frequency (f_{PLD}, Hz)	Number of pluses (N_{pulses})
3	650	35	10	300	10	30000
	-	-	-	-	15	-
	-	-	-	-	5	-

Table 5.6.The PLD conditions for set 4 (only the laser energy E_{PLD} was changed).

Sample set	Growth temperature $(T_g, °C)$	Pressure $(P_{O2}, mTorr)$	Distance (D, cm)	Laser energy (E_{PLD}, mJ)	Laser Frequency (f_{PLD}, Hz)	Number of pluses (N_{pulses})
4	650	35	10	300	10	30000
	-	-	-	325	-	-
	-	-	-	350	-	-
	-	-	-	400	-	-

Table 5.7.The PLD conditions for set 5 (only the pressure P_{O2} was changed).

Sample set	Growth temperature $(T_g, °C)$	Pressure $(P_{O2}, mTorr)$	Distance (D, cm)	Laser energy (E_{PLD}, mJ)	Laser Frequency (f_{PLD}, Hz)	Number of pluses (N_{pulses})
5	650	35	10	300	10	30000
	-	15	-	-	-	-
	-	50	-	-	-	-
	-	75	-	-	-	-
	-	100	-	-	-	-

The morphology of the ZnO materials obtained was studied by using FEI Nova Nano microscope SEM in the KAUST core lab. SEM provides immediate feedback on the optimized conditions for the nanostructure production. All PL measurements were carried out at liquid nitrogen temperature (77 K), employing 325 nm laser (KIMMON Ltd.), with the power output of about 1.6 mW after the lens. For XRD, X-ray diffraction (XRD) θ-2θ scans with a scan step of 0.002 $^\circ\theta$ (Bruker D8 Discover, λ_{av}=1.5418 Å) were employed.

5.2. The Effects of Growth Conditions

In this section, the effects of changing each deposition parameter on the obtained ZnO structure are discussed.

a) The effect of target-substrate distance (D)

The findings of the previous studies conducted by our group indicate that changing the target-substrate distance D had a significant effect on the material structure and can transfer the material structure from 2-dimensional to 1-dimensional. In order to better understand this phenomenon, this project commenced by setting the experimental conditions to those reported in Table 5.3 (set 1). This enabled studying the material structure dependence on the distance D. The SEM images shown in Figure 5.1 provide a visual record of the effect of D on the Eu-doped ZnO (0.15 wt% Eu) material structure. As can be seen, the most optimal conditions that produce well-aligned vertical Eu-doped ZnO nanorods (of a hexagonal structure) are obtained at D = 10 cm, as shown in Figure 5.1 (d). To our knowledge, this is the first time that such a high quality Eu-doped ZnO structure was obtained at a target-substrate distance of 10 cm, using standard PLD without catalyst. Liu and colleagues had studied the effect of changing the distance from 5 to 12 cm on the ZnO nanorods morphology [39]. However, the results the authors reported were obtained by modifying the PLD chamber to work under very high oxygen pressure (the reported background pressure ranged from 25 to 20 Torr, whereas the standard PLD chamber works in the mTorr range) [49]. Their findings suggest that, by increasing the distance, nanorods of larger size can be obtained than is possible with continuous film-like structures [39]. In this study, increasing the distance to D = 12 cm was sufficient for producing higher density nanorods, as shown in Figure 5.1 (e). However, the nanorods obtained under these conditions are not well aligned and do not exhibit hexagonal structure, which may imply lower crystal quality.

Figure 5.1 (a) And 5.1 (d) depict the results obtained at D = 5 and 11.5 cm, respectively, both of which indicate the presence of random nanostructure features. On the other hand, Figures 5.1 (b) and 5.1 (c) (obtained at D = 9.5 cm and 10 cm, respectively) show quantum dot-like structure. Based on these findings, D = 10 cm was chosen as the optimal distance and served as the input value for the subsequent deposition set. Figure 5.1 (f) shows a 45° tilt view of the optimized sample that growth at 10 cm substrate-target distance. The length of the nanorods ranged between (83-86) mm and the width ranged between (60-69) mm.

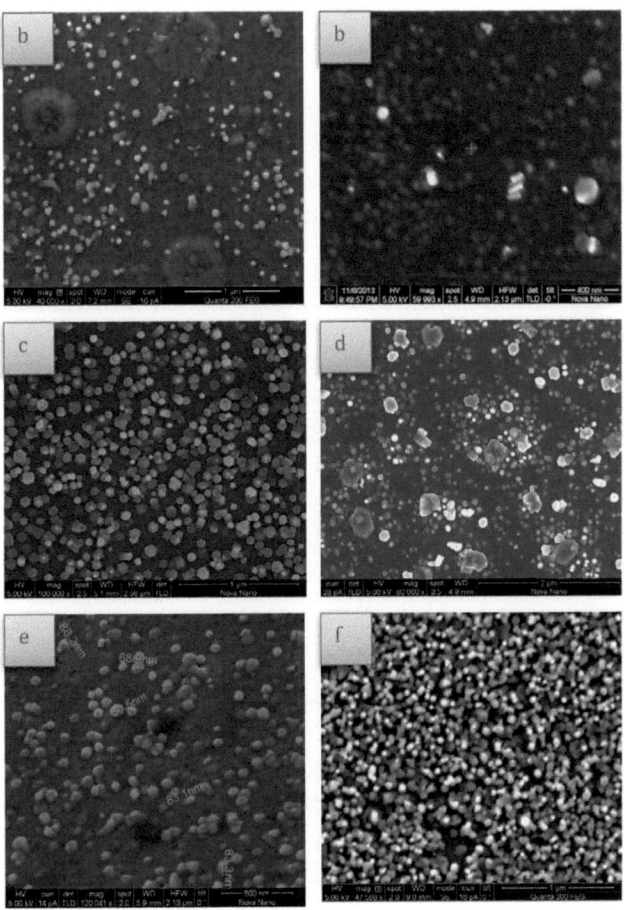

Figure 5.1. SEM top-view of Eu- doped ZnO (0.15 wt%) at various target-substrate distances (D): (a) D = 5 cm, (b) D = 9.5 cm, (c) D = 10 cm, (d) D = 11.5 cm, and (e) D = 12 cm, where (f) shows a 45° tilt view of the optimized conditions at 10 cm.

Figure 5.2 shows the PL spectrum of Eu-doped ZnO (0.15 wt%) nanostructures deposited at different distances D = 5, 10 and 12 cm, corresponding to samples shown in Figures 5.1 (a), 5.1 (c), and 5.1 (e), respectively. All samples were measured under the same PL conditions. The main peak centered at 372.9, 369.60 and 375.32 nm (for target-substrate distances of 5, 10 and 12 cm), respectively, is attributed to the ZnO band edge emission [64]. The redshift of the band edge emission obtained for the samples obtained at D = 5 and 12 cm is most likely due to the blue defect band overlapping with the band edge emission. On the other hand, the sample obtained with the optimized distance (D= 10 cm) shows a sharp ZnO band edge emission with no defect band. Moreover, the much narrower full width half maximum (FWHM), compared to two other samples (D = 5 and 12 cm), is indicative of a greater crystal quality, given that empirical evidence suggests that a broader band edge emission is associated with lower crystal quality. The two red sharp peaks at ~ 550 and 620 nm are due to Eu^{3+} emission, arising from $^5D_0 \rightarrow {}^7F_1$ and $^5D_0 \rightarrow {}^7F_2$ transitions, respectively [64]. These results may indicate that Eu^{3+} ions can be incorporated into the ZnO nanorods. The spectra of samples obtained with the distances of 5 cm and 12 cm show a broad defect band centered at 500 nm, which can be attributed to point defects, such as oxygen and zinc vacancies or interstitials [65]. No Eu^{3+} emission is observed in these samples, confirming low crystal quality. The Eu^{3+} emission is suppressed because the energy transferred from the ZnO host to Eu^{3+} ions is insufficient, due to the high density of defects [65]. Therefore, the PL measurements concur with the previously obtained SEM results, confirming that the sample grown at D = 10 cm exhibits the highest crystal quality.

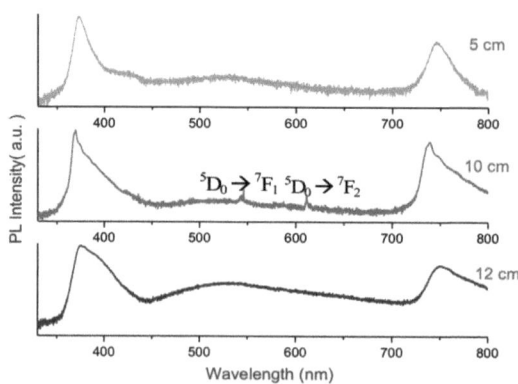

Figure 5.2. PL spectrum illustrates the substrate-target distance effects.

b) The effect of temperature (Tg)

In this set, as previously explained, the optimized conditions obtained in the previous set are used as input. In other words, the distance D = 10 cm remained constant, while the substrate temperature T_g changed in order to study its effects. The PLD deposition conditions pertaining to the Eu-doped ZnO (0.15 wt%) used in set 2 are shown in Table 5.4. As can be seen in the SEM images shown in Figure 5.3 (a-c), changing T_g had a significant effect on the structure of Eu-doped ZnO materials. Although the temperature was not changed significantly (in each step, 50°C increments were made), profound effects on the Eu ZnO structure were observed. For example, at 600°C, granular film was obtained, as shown in Figure 5.3 (a). On the other hand, the nanorod density decreased as the T_g increased, whereby film-like structures could not be observed at higher temperatures. This is evident in Figure 5.3 (b), which shows a separated vertically aligned nanorod structure obtained at T_g = 650°C, while very short nanorods of lower density (exhibiting hexagonal quantum dot-like structure) were obtained at T_g= 700°C, *as* shown in Figure 5.3(c). These results are in accordance with those reported in previous work examining T_g effects on the characteristics of the ZnO nanostructures [1, 2]. For example, Liu studied formation of poorly aligned nanorods under extreme conditions, created by modifying the PLD chamber to work at extreme oxygen pressure in a range of 0.25 to 20 Torr [39]. Liu reported that at low temperature (400°C) the film obtained had rough surface and a lump-like structure. However, by increasing T_g to 500°C, Liu was able to obtain nanorods of continuous morphology, capable of forming film-like structures. A further increase of T_g to 700°C resulted in isolated hexagonal crystals, enabling formation of nanorods. At 780°C, Liu reported that the nanorods lost their alignment and, once again, resumed the continuous film structure [39]. In a similar study, Henley et al. observed changes in nanorod structure as the temperature changed. They reported that the rods formed at Tg of 350°C had large diameter, thus resembling close-backed nanorods [66]. Thus, based on these findings, and the results obtained in the present study, it is evident that the nucleation process required for nanorod formation can occur at 600°C < T_g < 700°C [66]. This clearly indicates that the substrate temperature is the most important growth parameter.

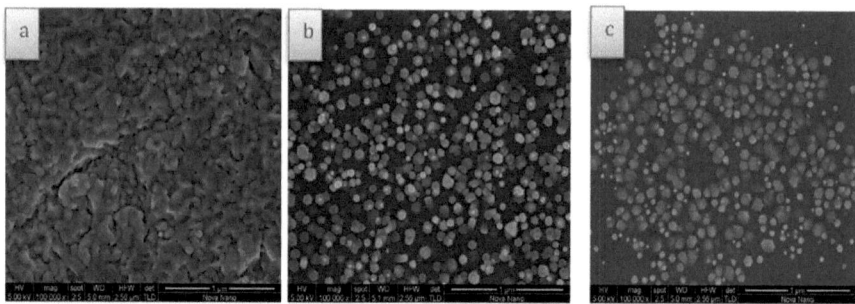

Figure 5.3. SEM images for set 2, corresponding to Eu-doped ZnO (0.15 wt% Eu): (a) film-like structure obtained at 600°C, (b) nanorods formed at 650°C and (c) quantum dot-like structure visible at 700°C.

In Figure 5.4, low temperature (77 K) PL spectra corresponding to the three T_g conditions, namely T_g= 600, 650 and 700°C, are shown. At each temperature, the band edge emission (at ~369 nm) is sharper, compared to those observed when distance D was changed. Moreover, no shift in the band edge emission can be seen as T_g increases. The PL spectra obtained for the samples grown at 600 and 700°C show two peaks in the band edge emission (369.23, 369.14 and 369.6 nm), which are due to free and bound exciton, respectively [35]. However, the spectrum obtained for the sample grown at 650°C (which corresponds to the optimized conditions), does not show the bound exciton peak. On the other hand, the red sharp emission of Eu^{3+} ions is shown at 610 nm [65] in this sample only. This result clearly indicates that energy is transferred from bound exciton of the host ZnO to Eu^{3+} intra-4f levels [14]. Moreover, the fact that compared to other conditions, no significant band defects were observed in the sample grown at 650°C indicates its higher crystal quality. On the other hand, the broad defect band centered at 550 nm is most pronounced for the film-like sample grown at 600°C. Finally, the FWHM for the sample deposited at 600°C is narrower (3.25 nm) than that measured for those grown at 650 and 700°C (3.98 and 3.53 nm, respectively). As the PL spectrum further confirms the conclusions derived from the SEM results, the deposition at 650°C is chosen as the optimal temperature condition. Therefore, the optimized conditions in set 2 are the same as those in set 1.

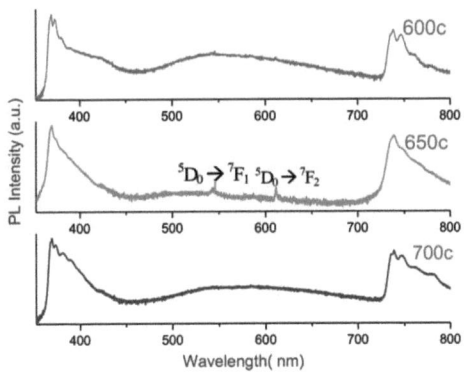

Figure 5.4. PL spectra of set 2 for Eu-doped ZnO (0.15 wt%) samples.

c) The effect of laser frequency (f_{PLD})

The aim of the sample set 3 is to investigate the effects of changing the deposition duration (f_{PLD}) on the formation of Eu ZnO nanostructures. The deposition conditions examined in this set are presented in Table 5.5. As can be seen in the SEM images shown in Figure 5.5 (a-c), modifying the f_{PLD} did not result in any significant changes in the structure of Eu ZnO (0.15 wt% Eu) nanorods. However, empirical evidence indicates that the pulse duration is an important determinant of the nanorod length and diameter. The three images presented in Figure 5.5 confirm that the length of the Eu ZnO nanorods decreases and their diameter increases as f_{PLD} decreases. For $f_{PLD} = 15$ Hz, the nanorod length ranges from 69 to 143 nm and its diameter equals 36.8 nm, as shown in Figure 5.5 (a). The nanorod length decreases significantly at 10 Hz, ranging between 54 and 98 nm, while the diameter increases (60.5-68.9 nm), as shown in Figure (5.5.b). Finally, when f_{PLD} is reduced to 5 Hz, as is clearly visible in Figure 5.5(c), no nanorods are formed, as quantum dot-like structure is observed, rather than the hexagonal nanostructure. The effect of f_{PLD} on the nanorod morphology was previously studied by Liu who found that increasing the f_{PLD} from 5 to 10 Hz resulted in larger rod size, which resembled a continuous film [39]. Hartanto et al. Also studied the effects of deposition time (10 to 300 s) on nanorod growth. The authors reported that at 10 s, only islands of thin film covered the substrate [67].

51

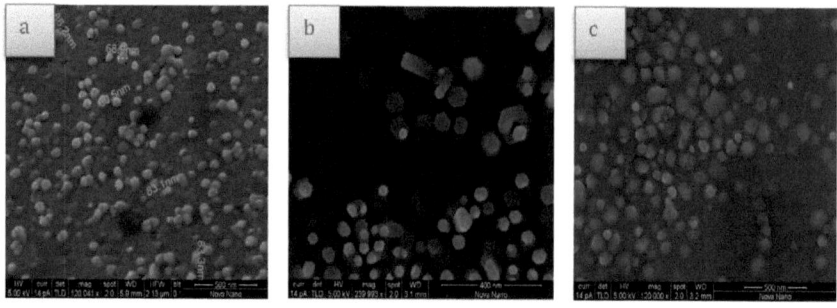

Figure 5.5.The effect of frequency f_{PLD} on Eu-doped ZnO (0.15%wt) nanostructure characteristics: (a) 15 Hz, (b) 10 Hz and (c) 5 Hz.

The low temperature PL spectra (77 K) corresponding to the samples obtained at 15, 10 and 5 Hz is shown in Figure 5.6. As can be seen, only the free exciton peaks are visible in all three cases. However, a blue shift in the ZnO band edge peak is observed as f_{PLD} decreases, which is centered at 373.3, 369.6 and 368.9 nm, for 15, 10 and 5Hz, respectively. This blue shift increases as the length of the nanorods increases and the diameter decreases, a phenomenon most likely due to carrier confinement effect [35]. It is known that the exciton feature experiences blue shift as the carrier confinement increases. The spectra presented in Figure 5.6 also indicate that the sample deposited at f_{PLD} = 15 Hz produces a broader band edge emission (FWHM = 8.73 nm) compared to the remaining two (3.67 and 4.12 nm, corresponding to 10 and 5 Hz, respectively). The broadening of the band edge emission can be due to the ZnO emission contribution from the ZnO layer underneath the nanorods, as the density of the nanorods produced at 15 Hz is lower than that obtained at 10 Hz. The defect band at 450 nm, which can be due to Zn_i, can be due to the emission contribution of the nanostructure, which starts to predominate as the density of the nanorods increases with f_{PLD} [35]. In addition, a broad defect band at 550 nm is observed for the sample deposited at 5 Hz. Based on these findings, it can be concluded that 15 Hz and 10 Hz yield similar results and either f_{PLD} can be chosen as the optimized condition. Thus, the decision to adopt a specific laser frequency is primarily driven by the desired nanorod size.

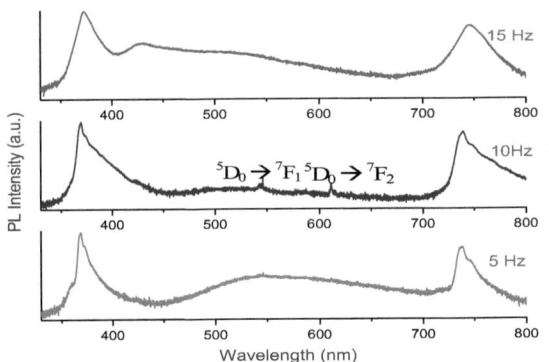

Figure 5.6. PL spectrum showing the effects of laser frequency on the optical properties of Eu- doped ZnO (0.15 wt%) nanorods.

d) The effect of laser energy (E_{PLD})

In the sample set 4, the effect of laser energy E_{PLD} on the formation of Eu ZnO nanorods is investigated. Previously, Ye Sen [68] studied the effect of the E_{PLD} on the ZnO nanorods at laser energy E_{PLD} = 15, 25, 45 and 80 mJ and at incident fluences F = 4.6, 11 and 20 Jcm^{-2}. The authors reported that at F= 4 Jcm^{-2}, nanorods of length =200 nm and diameter = 10 nm could be obtained. However, when the fluence energy increased to F = 11 Jcm^{-2}, the nanorod length reduced, while the diameter increased (ranging from 30 to 50 nm). The authors thus concluded that greater laser energy corresponded to shorter nanorod length, while increasing the diameter [68].

Table 5.6 shows deposition conditions used to produce the set 4 samples. The effect of laser energy on ZnO Eu (0.15 wt%) is shown in SEM images presented in Figure 5.7 (a-d). As can be seen, the sample deposited at 300 mJ (the laser energy that was used in the preceding sets) shows vertical well-aligned nanorods characterized by hexagonal structure Figure 5.7 (a). As the E_{PLD} increases to 350 mJ, while the structure quality of the vertical nanorods is maintained, their density increases Figure 5.7 (c). As the E_{PLD} increases to 400 mJ, the nanorod diameter decreases figure 5.7 (d).

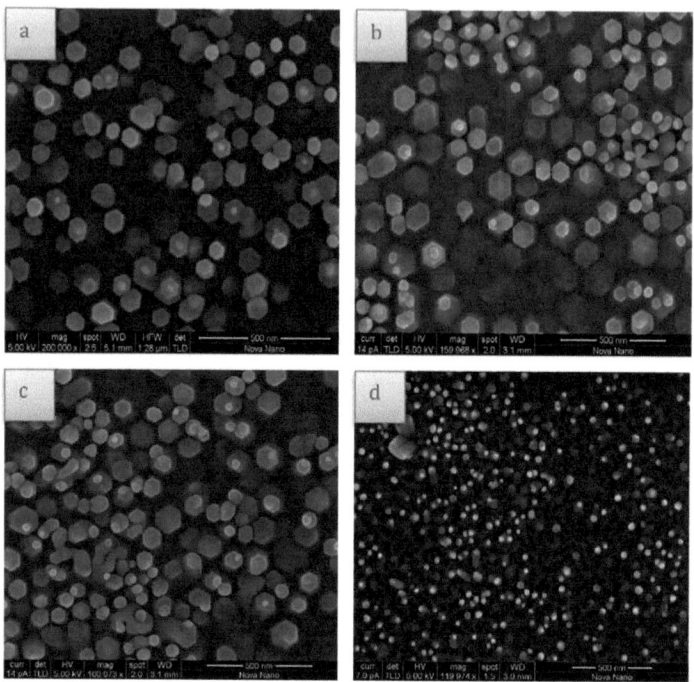

Figure 5.7. SEM images for Eu-doped ZnO (0.15 wt% Eu) nanorods deposited at different laser energy E_{PLD} (a) 300 mJ, (b) 325 mJ, (c) 350 mJ and (d) 400 mJ.

Figure 5.8 shows the PL spectra corresponding to these samples, obtained at low temperature (77 K). As can be seen, the ZnO band edge peaks are located at 376.2, 373.8, 373.8 and 369.7 nm, for the samples grown at E_{PLD}= 400, 350, 325 and 300 mJ, respectively. The obvious shift of the band edge peak can be due the broadening of the peak, as the FWHM decreases with the increase in E_{PLD}. More specifically, the FWHM of the samples deposited at 400, 350, 325 and 300 mJ is measured at 11, 15.87, 10.52 and 3.99 nm, respectively. This variation may imply the emission contribution from the ZnO layer underneath the nanorods. The intensity of the defect band also increases as the E_{PLD} increase, which is in agreement with the band edge trend. The Eu emission is only observed in the sample deposited at 300 mJ.

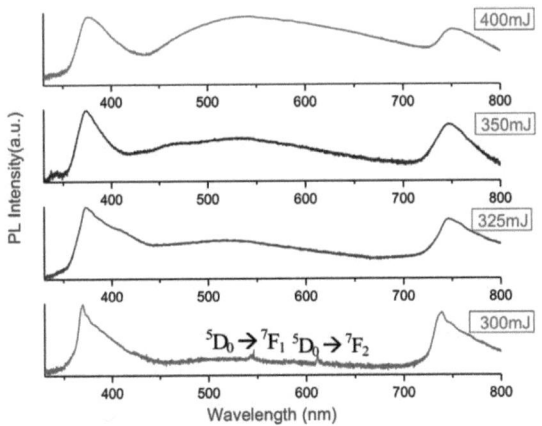

Figure 5.8.PL spectrum of the effect of the laser energy E_{PLD} on the optical properties of Eu- doped ZnO (0.15 wt%) nanorods.

e) The effect of pressure (P_{o2})

In the sample set 5, the objective is to investigate the effects of changing the partial oxygen pressure (Po₂) during PLD deposition. Table 5.7 shows the growth conditions used in this set, while the SEM images presented in Figures 5.9 (a-e) provide visual evidence of the effects of different P_{O_2} conditions on the formation of the ZnO Eu (0.15 wt%) nanorods. As can be seen, increasing the pressure results in the formation of nanorods of greater density [39]. This outcome is likely due to the greater oxygen pressure producing rougher samples, which may lead to the formation of nano-nucleation centers and thus higher-density nanorods [39]. Zuniga-preez et al. [69] studied the effect of P_{o2} on high pressure pulsed laser energy, as well as the structural properties of ZnO nanorods. In their work, the authors observed that at low pressure, the nanorod growth rate was reduced, while ZnO nanowires started to form at pressures near 200 mbar. Thus, they concluded that the pressure and the number of pluses play a significant role in the nanowire growth control. In addition, Liu modified the PLD chamber to work at a significant deposition pressure (1-25 Torr) to find that contentious film can be produced at lower P_{o2}, while separated nanorods can be formed at significantly higher pressure (25 Torr) [39]. Finally, increasing the P_{o2} above 20 Torr leads to larger nanorods, which may form continuous film. These results indicate that determining the optimal P_{o2} is significant for controlling the nanorod density. However, the exact P_{o2} values will depend on the application.

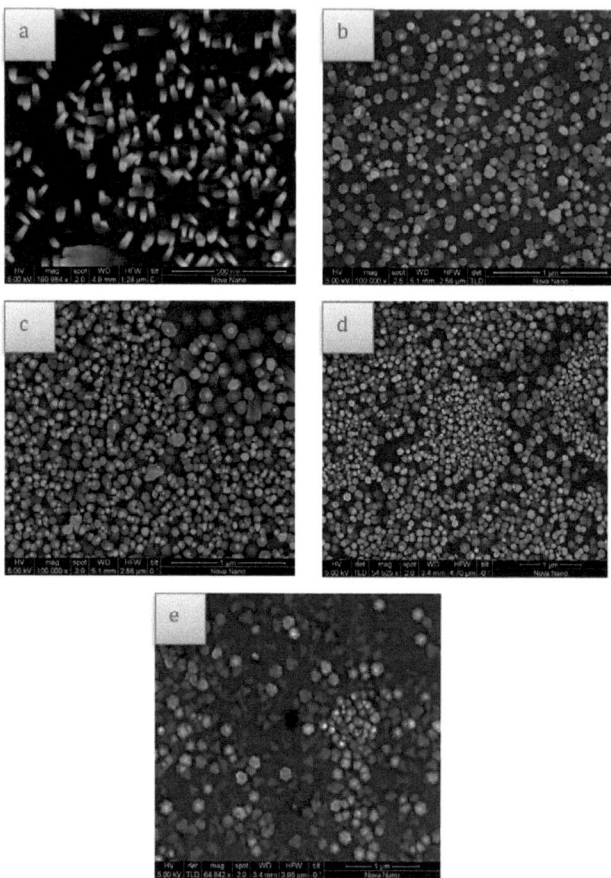

Figure 5.9. SEM images of Eu-doped ZnO (0.15 wt%) samples grown at different P_{o2}: (a) 15, (b) 35, (c) 50, (d) 75 and (e) 100 mTorr.

Figure 5.10 shows the low temperature (77 K) PL spectra corresponding to the samples grown at different P_{o2}. As can be seen, at 100, 75, 50, 35 and 15 mTorr, the band edge peak of ZnO is located at 367.2, 368.9, 373.3, 369.8 and 375.9 nm, respectively. Moreover, the samples grown at 100, 75 and 35 mTorr show sharper ZnO band edge peaks compared to those obtained at other P_{o2}. The defect band is only found at 50 and 75 mTorr pressure conditions, as no significant defect band is visible at 15, 35 and 100 mTorr, indicating high crystal quality. Based on the SEM and PL measurements, it can be concluded that 35, 75 and 100 mTorr are the best conditions for nanorod growth. The FWHM for the samples deposited at 100, 75, 50, 35 and 15 is 3.1, 3.3, 15, 3.99 and 21.06 nm, respectively.

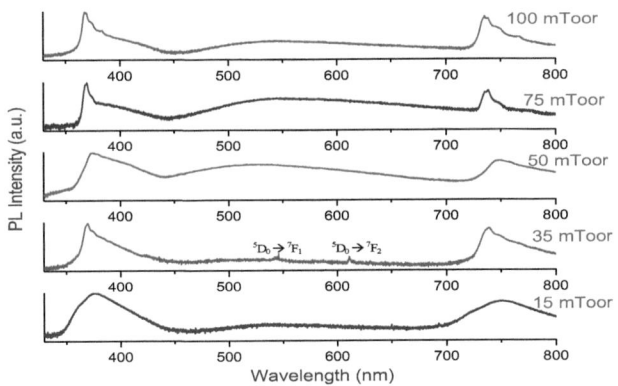

Figure 5.10.PL spectra of Eu-doped ZnO (0.15 wt% Eu) nanorods grown at 15, 30, 50, 75 and 100 mTorr.

f) The effect of Eu concentration

In this section, the effect of changing the Eu concentration on the formation ZnO nanorods is discussed. Figure 5.11 (a-c) shows the SEM images of the samples grown at different Eu concentrations while keeping the pressure at 35 mTorr. The nanorods grown under the optimized conditions used in the previous sections are presented in Figure 5.11(b). As can be seen, when undoped ZnO samples are grown under similar PLD conditions, no nanorods, Figure 5.11(a), are formed, as only a lump-like surface is visible. On the other hand, samples grown at higher Eu concentration (0.2 wt%) show continuous film structure. This finding suggests that nano-nucleation density increases with increasing amounts of Eu dopants, as their presence assists in increasing the nanorod density, leading to the formation of a continuous film.

Figure 5.12 shows the effect of P_{o2} for these three targets. As P_{o2} increases to 50 mTorr, a different response is observed for each material. Undoped ZnO is shown in Figure 5.12 (a); Eu-doped ZnO (0.15 wt %Eu) is depicted in Figure 5.12 (b); and Eu-doped ZnO (0.15 wt %Eu) is presented in Figure 5.12 (c). These results are compared to those obtained at 35 mTorr. As can be seen, samples with higher Eu concentrations are characterized by film-like structure, with grains smaller than those observed in undoped ZnO.

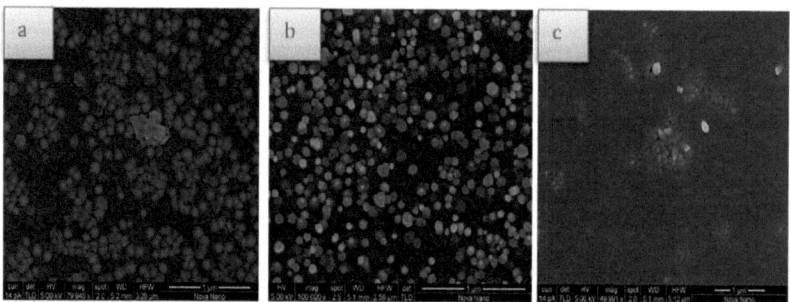

Figure 5.11.SEM images for various ZnO materials: (a) undoped ZnO, (b) Eu-doped ZnO (0.15 at %) and (c) Eu-doped ZnO (0.2 at %) at 35 mTorr.

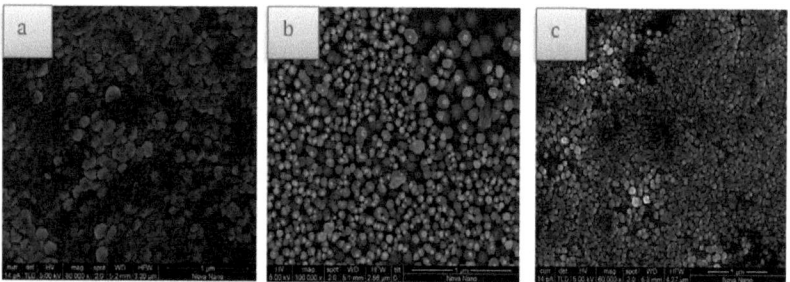

Figure 5.12. The structures obtained at 50 mTorr are shown below: (a) pure ZnO oxide (b) Eu-doped ZnO (0.15 at %) and (c) Eu-doped ZnO (0.2 at %wt).

In line with the previous sets, low temperature (77 K) PL spectra of the samples grown at 35 mTorr were obtained and are presented in Figure 5.13. Broader band edge emission is observed for undoped ZnO and Eu ZnO (0.2 at % Eu), compared to Eu ZnO (0.15 wt % Eu) nanorods. However, a weak defect band is observed in all three samples. The PL spectra corresponding to these samples grown at 50 mTorr are shown in Figure 5.14. It is evident that, for the samples grown under greater pressure, weaker and broader band edge peaks are observed compared to those grown at 35 mTorr. This finding indicates that the materials obtained at 50 mTorr are of a lower crystal quality, which is in agreement with the results of SEM imaging. Therefore, it can be concluded that (0.15 wt%) is the optimal Eu concentration, when combined with the previously chosen PLD conditions and the target preparation adopted in this study.

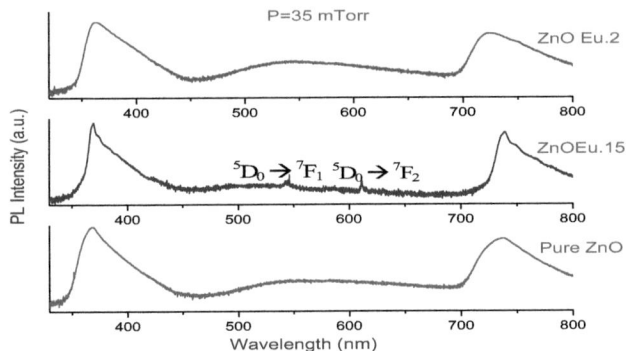

Figure 5.13. The PL for ZnO nanorods at different Eu concentrations and at 35mTorr.

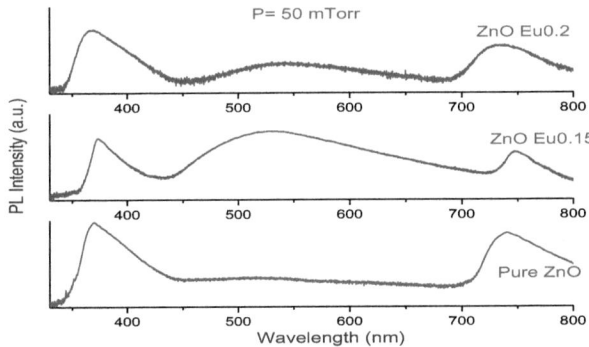

Figure 5.14. The PL for ZnO with different Eu concentrations at 50 mTorr.

In conclusion, based on the experimental sets in which a wide range of PLD and target preparation conditions was explored, D = 10 mm, T_g = 650 °C, E_{PLD} = 300 mJ-350 mJ, and f_{PLD} = 10 Hz are the most optimal conditions for nanorod growth (using 30,000 pulses).

5.3 XRD results

Finally, XRD measurements were carried out in order to estimate the growth orientation of the materials. Although TEM measurements are necessary for determining the exact structure of the ZnO nanostructure, this was not possible in this work due to the limited laboratory access during the period of the four months of my research. Thus, only XRD was performed, whereby pure ZnO and Eu-doped ZnO nanorods (with different Eu concentrations) were examined. These were grown on *a*-sapphire at 35 mTorr and were dominated by the (002) reflection. Figure 5.15 shows 2-theta scans corresponding to the optimized undoped and Eu-doped ZnO nanostructures for different Eu compositions. The (002) peak is located at 34.77°, 34.7° and 34.6° for Eu-doped ZnO samples (corresponding to 0, 0.2 and 0.15 at% Eu), respectively. All other peaks correspond to those of the substrate (sapphire Al_2O_3). Therefore, these ZnO nanostructures are highly c-axis oriented. The ZnO (002) peak is sharper for undoped ZnO compared to that obtained for doped ZnO, which indicates that the former produces better crystal quality. This outcome likely arises because of Eu incorporation (i.e., replacement of Zn by Eu atoms in the lattice) and may cause crystal distortion due to its much larger atom size compared to that of Zn [70] .

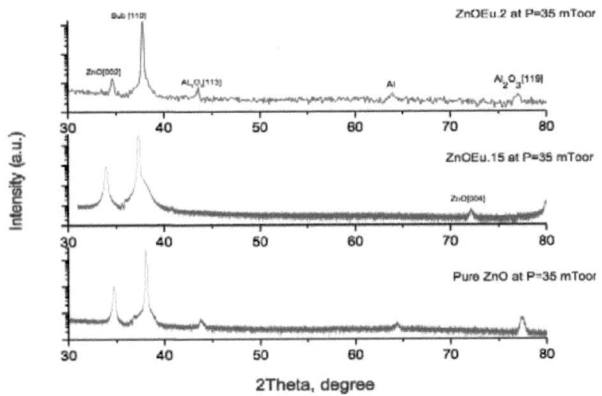

Figure 5.15. 2-theta scan of undoped ZnO and Eu-doped ZnO (0.15, 0.2 wt%) at 35 mTorr

Chapter 6
Discussion

In this chapter, a brief overview of the work on ZnO nanorod growth by PLD without catalysts is presented. This is followed by a discussion of the effects of different conditions adopted during the PLD process on the morphology and optical properties of the ZnO nanorods. In particular, the main observations that arose during the experimental work conducted in this study are discussed and compared to the previous findings.

6.1 Introduction

Mitsuhisa Kawakami reported that he successfully synthesized the ZnO nanorods by nanoparticle assisted PLD without catalyst. In this study, the growth conditions included extremely high oxygen pressure of around 5-10 Torr **(that can be achieved by modifying PLD system, as the standard PLD system works in mTorr range.) [71]**, With the very small target-substrate distance of 2 cm. Nanorods have c-axis orientation, and are typically grown to have a diameter of 300 nm and a length of 6 μm. Moreover, the PL spectrum obtained at room temperature demonstrated an intense, narrow UV emission band and only very weak visible emission, indicating that the quality of ZnO nanorods was very high [71]. Hartanto reported on the work on the nanoparticles formed in the laser ablation plume, which was transported onto the substrate to grow the nanorods [67]. The nanorods obtained by this approach had 300 nm diameters and a length of 6 μm [67]. Liu reported that c-axis oriented nanorods grown on both sapphire and silicon substrates by PLD techniques without catalyst were affected by the choice of the deposition parameters [39]. They have demonstrated that, by adjusting the pressure and the substrate temperature in the range of 5-20 Torr and 550-700°C respectively, the size of the nanorods could be successfully controlled [39].

Perez reported that he and his colleagues had successfully grown tilted (30°) ordered ZnO nanowires on m-plane sapphire by catalyst-free PLD [69]. The authors proposed that the total number of pulses and the oxygen partial pressure could have a significant effect on the nanowire length and diameter. In this work, the optical properties of the nanowires were improved by modifying the partial pressure [69].

Guo and his group reported that they had successfully synthesized vertically aligned ZnO nanonails on annealed sapphire substrate by nanoparticle-assisted pulsed-laser deposition (NAPLD) at comparatively high argon gas pressure without catalyst [72]. Their XRD and SEM results showed that each of the uniquely shaped

ZnO nanonails consisted of a hexagonal rod shaped "rot". Furthermore, when the optical properties of the ZnO nanonails were examined, it was evident that at room temperature, they exhibited a strong ultraviolet (UV) emission at 390 nm, accompanied by negligible visible emission. These findings indicated that in highly oriented ZnO nanonails, the concentration of oxygen vacancies is very low [72].

Ryo Nishimura published the results of a study in which ZnO nanorod arrays were successfully grown on a Si (100) substrate by PLD [11]. The growth was performed by first dispersing the ZnO powder on the Si substrate, before annealing it to form seeds for nanorod growth. In this work, two processes (annealing and off-axis PLD) were employed and their differences analyzed. In this context, the term "off-axis" indicates that the substrate is positioned away from the position typically used in the PLD approach [11].

Varanasi reported that ZnO nanorods were grown on Si and sapphire substrate, in a modified PLD system with a large chamber, with or without the seed layer [73]. A wide variety of seed materials were used, including Au, Cr and BaSrTiO$_3$ (BST) and the results obtained in each case were compared in order to identify the most optimal growth conditions. The author reported that the choice of the substrate and seed material affected the alignment of ZnO nanorods. More specifically, nanorods grown on sapphire showed the best alignment, when Cr or BST were used as seed layers [73]. Therefore, all of these previous studies obtained ZnO nanorods by catalyst, nanoparticles seeding or modifying the PLD chamber.

6.2 Discussion

In the present study, SEM imaging revealed that changing the temperature significantly affected the morphology of the ZnO nanorods produced. At 650°C, for example, isolated hexagonal crystal nanorods could be observed. However, above and below 650°C ZnO nanorods lost their ordered alignment and their structure resumed the continuous film quality. Similarly, increasing the pressure above 35 mTorr resulted in greater distribution of the nanorods. As a result, the nanorods aggregated together to form continuous film again. By extending the growth rate (deposition duration), it was possible to obtain nanorods of greater length. It is noteworthy that substrate distance D has a significant effect on the resultant nanorod structure. The experiments performed while changing P, T, and f indicated that these factors also played a role in nanorod formation, albeit to a lesser degree. For PLD, the optimum distance for obtaining high quality nanorods was found to be 30 mm, as film was the resulting structure at all other distances. As a part of the experimental process, the desired plume length was obtained by adjusting (D) according to the variation in the pressure. However, the findings obtained confirmed that distance was not affected by temperature [39].

We believe that these nanostructures occurred though nano-nucleation. The main effect is the surface energy of the grown materials, the interface and the substrates, as shown in chapter 3. Many factors can control the surface energy to produce the nanorod nucleation and growth of the crystalline films in the PLD. However, empirical evidence indicates that the temperature, material type, substrate properties and density, and the degree and energy of the ionization species in the plume have the greatest effect [39]. The substrate temperature and supersaturation are the main thermodynamic factors implicit in the growth mechanism, whereby the latter depends on the pressure and the laser energy. Low deposition rate is found to result in the creation of large nuclei. Consequently, isolated patches of the film that coalesce together are created on the substrate. As the supersaturation increases, the nucleus size tends to decrease, until it reaches an atomic diameter. This results in a 2-D layer. Excessive supersaturation is not favorable for growing nanorods because when the substrate is incompletely wetted, the layer-by-layer nucleation tends to occur [39]. In their work, Liu and Ong [74] reported that for producing discontinuous film, low deposition rate and high substrate temperature are required. This is in good agreement with the results obtained in the present study, where continuous film was obtained at high temperature, due to the unbalanced liquid/solid ratio. For the same reason, continuous film was produced at low deposition rate [39].

It should be noted that each surface of wurtzite ZnO consists of several planes, such as $(10\bar{1}0)$, $(11\bar{2}0)$, $(10\bar{1}1)$ and (0001), all characterized by different surface energy. According to the available data, the highest surface energy is associated with (0001), while that corresponding to $(11\bar{2}0)$ lower than [7]. Based on these findings, it is evident that a substantial surface energy gain can be obtained from producing nanorods, rather than thin films. Moreover, due to the reduced surface area of ZnO (0001), the total surface energy is minimized [7]. Using substrate with high isotropic surface energy can enable 1D ZnO nanorod growth because, during the initial growth, ZnO nuclei can occur randomly across the entire substrate surface. As a result, the nanorods characterized by a reduction in the surface formation energy are produced. Using highly anisotropic surface energy allows the growth mode and morphology of ZnO to be successfully controlled [7].

6.3 Conclusion

In summary, as was shown in the work reported thus far, the PLD parameters must be carefully controlled in order to obtain c-axis oriented ZnO nanorods on sapphire substrate, without the use of catalyst. The experiments conducted in order to identify the optimal growth conditions confirmed that by adjusting the target-substrate distance, temperature, laser energy and deposition duration, the nanorod size could be successfully controlled. Most importantly, the PL properties reflect the quality of the ZnO nanorods.

In the future, this work will continue by studying the effects of the various substrate types on the characteristics of the ZnO nanorods obtained. Moreover, work on stabilizing the energy of the laser beam for all the conditions will be performed by changing the glass filter periodically (as and when needed) and measuring the energy of the beam before each deposition.

References

1. Takahashi, K., Yoshikawa, A. and Sandhu, A. (2007). *Wide bandgap semiconductors*. 1st ed. Berlin: Springer.

2. Djurisic, A. and Leung, Y. (2006). Optical properties of ZnO nanostructures. *Small*, 2(8-9), pp. 944-961.

3. Jagadish, C. and Pearton, S. (2006). *Zinc oxide bulk, thin films and nanostructures*. 1st ed. Amsterdam: Elsevier.

4. Yogamalar, R. and Bose, A. (2013). Synthesis, dopant study and device fabrication of zinc oxide nanostructures: mini review. *Progress in Nanotechnology and Nanomaterials*, 2(1), pp. 1-20.

5. Jagadish, C. and Pearton, S. (2006). *Zinc oxide bulk, thin films and nanostructures*. 1st ed. Amsterdam: Elsevier.

6. Litton, C., Reynolds, D. and Collins, T. (2011). *Zinc oxide materials for electronic and optoelectronic device applications*. 1st ed. Hoboken: Wiley.

7. Yi, G. (2012). *Semiconductor nanostructures for optoelectronic devices*. 1st ed. Heidelberg: Springer.

8. Stwertka, A. (2002). *A guide to the elements*. 1st ed. New York: Oxford University Press.

9. Yi, G. (2012). *Semiconductor nanostructures for optoelectronic devices*. 1st ed. Heidelberg: Springer.

10. Morkoc, H. and Ozgur, U. (2009). *Zinc oxide fundamentals, materials and device technology*. 1st ed. Weinheim: Wiley-VCH.

11. Nishimura, R., Sakano, T., Okato, T., Saiki, T. and Obara, M. (2008). Catalyst-free growth of high-quality ZnO nanorods on Si (100) substrate by two-step, off-axis pulsed-laser deposition. *Japanese Journal Of Applied Physics*, 47(6R), pp. 4799

12. Behera, J. K. (2010). *Synthesis and characterization of ZnO nano-particles*. Master Thesis, National Institute of Technology, India.

13. Konar, A., Verma, A., Fang, T., Zhao, P., Jana, R. and Jena, D. (2012). Charge transport in non-polar and semi-polar semiconductor III-V nitride heterostructures. *Semiconductor Science and Technology*, 27(2), pp. 024018.

14. Ishizumi, A. and Kanemitsu, Y. (2005). Structural and luminescence properties of Eu-doped ZnO nanorods fabricated by a microemulsion method. *Applied Physics Letters*, 86(25), pp. 253106.

15. Carter, C. and Norton, M. (2007). *Ceramic materials*. 1st ed. New York: Springer.

16. Klingshirn, C., Waag, A., Hoffmann, A. and Geurts, J. (2010). *Zinc oxide from fundamental properties towards novel applications*. 1st ed. Heidelberg: Springer.

17. Ayers, J. (2007). *Heteroepitaxy of semiconductors*. 1st ed. Boca Raton: CRC Press.

18. Jiles, D. (1994). *Introduction to the electronic properties of materials*. 1st ed. London: Chapman and Hall.

19. Janotti, A. and Van de Walle, C. (2009). Fundamentals of zinc oxide as a semiconductor. *Reports on Progress in Physics*, 72(12), pp. 126501.

20. Pierret, R. (1987). *Advanced semiconductor fundamentals*. 1st ed. Reading, Mass.: Addison-Wesley.

21. Meyer, B., Alves, H., Hofmann, D., Kriegseis, W., Forster, D., Bertram, F., Christen, J., Hoffmann, A., Strasburg, M. and Dworzak, M. (2004). Bound exciton and donor-acceptor pair recombinations in ZnO. *Physica Status Solidi (b)*, 241(2), pp. 231-260.

22. Feng, Z. (2013). *Handbook of zinc oxide and related materials*. 1st ed. Boca Raton: CRC Press.

23. Ishizumi, A., Taguchi, Y., Yamamoto, A. and Kanemitsu, Y. (2005). Luminescence properties of ZnO and Eu^{+3} doped ZnO nanorods. *Thin Solid Films,* 486(1-2), pp. 50-52.

24. Gfroerer, T. H. (2000). *Photoluminescence in analysis of surface and interface.* 1st ed. New York: Wiley.

25. Shimada, R., Urban, B., Sharma, M., Singh, A., Avrutin, V., Morkoc, H. and Neogi, A. (2012). Energy transfer in ZnO-anthracene hybrid structure. *Optical Materials Express,* 2(5), pp. 526-533.

26. Rodnyi, P. A. and Khodyuk, I. V. (2011). Optical and luminescence properties of zinc oxide (review). *Optics and Spectroscopy,* 111(5), pp. 776-785.

27. Aoki, T. (2012). *Photoluminescence spectroscopy: characterization of materials.* 1st ed. New York: Wiley.

28. Kumar, C. (2013). *UV-VIS and photoluminescence spectroscopy for nanomaterials characterization.* 1st ed. Berlin: Springer.

29. Roqan, I. (2014). *MSE 302 Electronic Properties of Materials.* Lecture, King Abdullah University of Science and Technology, Thuwal.

30. Sze, S. and Ng, K. K. (2006). *Physics of semiconductor devices.* 3rd ed. Hoboken: John Wiley & Sons.

31. McCluskey, M. and Haller, E. (2012). *Dopants and defects in semiconductors.* 1st ed. Boca Raton, FL: Taylor and Francis.

32. Djurisic, A. and Leung, Y. (2006). Optical properties of ZnO nanostructures. *Small,* 2(8-9), pp. 944-961.

33. Ozgur, U. Y. I., Alivov, C., Liu, A., Teke, M. A., Reshchikov, S., Dogan, V., Avrutin, S., Cho, J. and Morkoc, H. (2005). A comprehensive review of ZnO materials and devices. *Journal of Applied Physics,* 98(4), pp. 041301.

34. Ahmed, S., Szymanski, P., El-Nadi, L. and El-Sayed, M. (2014). Energy-transfer efficiency in Eu-doped ZnO thin films: the effects of oxidative annealing on the dynamics and the intermediate defect states. *ACS Applied Materials & Interfaces,* 6(3), pp. 1765-1772

35. Bergman, L. and McHale, J. (2012). *Handbook of luminescent semiconductor materials.* 1st ed. Boca Raton, FL: CRC Press.

36. Top- target. Com, (2014). Rare earth materials. [Online]. Available at: http: // www. http://www.top-targets.com/rare-earth-materials. [Access 8 Aug.2014].

37. Chen, L., Hu, H. and Xiong, Z. (2013). A role of Eu-doping on electronic structure and optical properties of ZnO from first-principles. *Applied Physics Frontier,* 1(2), pp. 22-26.

38. O'Donnell, K. and Dierolf, V. (2010). *Rare earth doped III-nitrides for optoelectronic and spintronic applications.* 1st ed. Dordrecht: Springer.

39. Liu, Z. and Ong, C. (2007). Synthesis and size control of ZnO nanorods by conventional pulsed-laser deposition without catalyst. *Materials Letters,* 61(16), pp. 3329-3333.

40. Polsongkram, D., Chamninok, P., Pukird, S., Chow, L., Lupan, O., Chai, G., Khallaf, H., Park, S. and Schulte, A. (2008). Effect of synthesis conditions on the growth of ZnO nanorods via hydrothermal method. *Physica B: Condensed Matter,* 403(19), pp. 3713-3717.

41. Cao, G. (2004). *Nanostructures and nanomaterials: synthesis, properties, and applications.* 1st ed. London: Imperial College Press.

42. Fons, P., Iwata, K., Yamada, A., Matsubara, K., Niki, S., Nakahara, K., Tanabe, T. and Takasu, H. (2000). Uniaxial locked epitaxy of ZnO on the face of sapphire. *Applied Physics Letters,* 77(12), pp. 1801-1803.

43. Cao, G. and Wang, Y. (2011). *Nanostructures and nanomaterials: synthesis, properties, and applications.* 2nd ed. New Jersey: World Scientific.

44. Machlin, E. (2005). *Materials science in microelectronics.* 1st ed. Amsterdam: Elsevier.

45. Kirby, K. (2008). *Processing of sapphire surface for semiconductor device applications.* Master Thesis, The Pennsylvania State University, Pennsylvania.

46. Kanel, G., Nellis, W., Savinykh, A., Razorenov, S. and Rajendran, A. (2009). Response of seven crystallographic orientations of sapphire crystals to shock stresses of 16-86 GPa. *Journal of Applied Physics,* 106(4), pp. 043524.

47. Wybourne, B. (1966). Energy levels of trivalent gadolinium and ionic contributions to the ground state splitting. *Physical Review,* 148(1), pp. 317-327.

48. Li, Z., Hu, Z., Liu, F., Hang, H., Zhang, X., Wang, Y., Jiang, L., Yin, P. and Guo, L. (2012). Lateral growth and optical properties of ZnO microcrystal on sapphire substrate. *Optical Materials,* 34(11), pp. 1908-1912.

49. Yi, G., Wang, C. and Park, W. (2005). ZnO nanorods: synthesis, characterization and applications. *Semiconductor Science and Technology,* 20(4), pp. 22.

50. Aliofkhazraei, M., and Rouhaghdam, A. (2010). *Fabrication of nanostructures by plasma electrolysis.* 1st ed. Weinheim: Wiley-VCH.

51. Wang, Z. L. (2004). Zinc oxide nanostructure: growth, properties and applications. *Journal of Physics: Condens Matter,* 16(4), pp. R829-858.

52. Eason, R. (2007). *Pulsed laser deposition of thin films.* 1st ed. Hoboken: John Wiley & Sons.

53. Schou, J. (2009). Physical aspects of the pulsed laser deposition technique: the stoichiometric transfer of material from target to film. *Applied Surface Science,* 255(10), pp. 5191-5198.

54. Lin, Y. H. (2009). *Structure and properties of transparent conductive ZnO films grown by pulsed laser deposition (PLD).* Master Thesis, University of Birmingham, Birmingham.

55. Mohmood, A. and Rouhaghdam, A. (2011). *Fabrication of nanostructures by plasma electrolysis.* 1st ed. Hoboken: John Wiley and Sons.

56. Leng, Y. (2008). *Materials characterization: introduction to microscopic and spectroscopic methods.* 1st ed. Hoboken. Wiley-VCH.

57. Williams, D. and Carter, C. (2009). *Transmission electron microscopy.* 1st ed. New York: Springer.

58. Driggers, R. (2003). *Encyclopedia of optical engineering.* 1st ed. New York: CRC Press.

59. Khursheed, A. (2011). *Scanning electron microscope optics and spectrometers.* 1st ed. Singapore: World Scientific.

60. Waseda, Y., Matsubara, E. and Shinoda, K. (2011). *X-ray diffraction crystallography: Introduction, Examples and Solved Problems.* 1st ed. Berlin: Springer.

61. Claflin, B., Look, D., Park, S. and Cantwell, G. (2006). Persistent n-type photoconductivity in p-type ZnO. *Journal of Crystal Growth,* 287(1), pp. 16-22.

62. Baras, A. (2011). *Sturucral and magnetic properties of Mn doped ZnO thin film deposited by pulsed laser deposition.* Master Thesis, King Abdullah University of Science and Technology, Thuwal.

63. Tan, Y., Fang, Z., Chen, W. and He, P. (2011). Structural, optical and magnetic properties of Eu-doped ZnO films. *Journal of Alloys and Compounds,* 509(21), pp. 6321-6324.

64. Yang, Y., Feng, Y., Zhu, H. and Yang, G. (2010). Growth, structure, and cathodoluminescence of Eu-doped ZnO nanowires prepared by high-temperature and high-pressure pulsed-laser deposition. *Journal of Applied Physics,* 107(5), pp. 053502.

65. Henley, S. J., Ashfold, M. N., Nicholl, D. P., Wheatley, P. and Cherns, D. (2004). Controlling the size and alignment of ZnO microrods using ZnO thin film templates deposited by pulsed laser deposition. *Applied Physics A,* 79(4-6), pp. 1169-1173.

66. Hartanto, A. B., Ning, X., Nakata, Y. and Okada, T. (2004). Growth mechanism of ZnO nanorods from nanoparticles formed in a laser ablation plume. *Applied Physics*, 78(3), pp. 299-301.

67. Sun, Y., Doherty, R., Warren, J. and Ashfold, M. (2007). Effect of incident fluence on the growth of ZnO nanorods by pulsed excimer laser deposition. *Chemical Physics Letters*, 447(4), pp. 257-262.

68. Zuniga-Perez, J., Rahm, A., Czekalla, C., Lenzner, J., Lorenz, M. and Grundmann, M. (2007). Ordered growth of tilted ZnO nanowires: morphological, structural and optical characterization. *Nanotechnology*, 18(19), pp. 195303.

69. Yoon, H., Wu, J., Min, J., Lee, J., Ju, J. and Kim, Y. (2012). Magnetic and optical properties of monosized Eu-doped ZnO nanocrystals from nanoemulsion. *Journal of Applied Physics*, 111(7), pp. 07523.

70. Kawakami, M., Hartanto, A., Nakata, Y. and Okada, T. (2003). Synthesis of ZnO nanorods by nanoparticle assisted pulsed-laser deposition. *Japanese Journal of Applied Physics*, 42(1A), p.33.

71. Guo, R., Nishimura, J., Ueda, M., Higashihata, M., Nakamura, D. and Okada, T. (2007). Vertically aligned growth of ZnO nanonails by nanoparticle-assisted pulsed-laser ablation deposition. *Applied Physics A*, 89(1), pp. 141-144.

72. Varanasi, C. V., leedy, K. D., Tomich, D. H., Subramanyam, G. and look, D. C. (2009). Improved photoluminescence of vertically aligned ZnO nanorods grown on BaSrTiO₃ by pulsed laser deposition. *Nanotechnology*, 20(38), pp. 1-5.

73. Metev, S. and Veiko, V. (1998). *Laser-assisted microtechnology*. 1st ed. Berlin: Springer.

I want morebooks!

Buy your books fast and straightforward online - at one of the world's fastest growing online book stores! Environmentally sound due to Print-on-Demand technologies.

Buy your books online at
www.get-morebooks.com

Kaufen Sie Ihre Bücher schnell und unkompliziert online – auf einer der am schnellsten wachsenden Buchhandelsplattformen weltweit! Dank Print-On-Demand umwelt- und ressourcenschonend produziert.

Bücher schneller online kaufen
www.morebooks.de

VDM Verlagsservicegesellschaft mbH
Heinrich-Böcking-Str. 6-8   info@vdm-vsg.de
D - 66121 Saarbrücken   Telefax: +49 681 93 81 567-9   www.vdm-vsg.de

MIX
Papier aus verantwortungsvollen Quellen
Paper from responsible sources
FSC® C105338

Printed by Books on Demand GmbH, Norderstedt / Germany